ASHEVILLE-BUNCOMBE TECHNICAL INSTITUTE

DISCARDED

NOV 20 2024

An Introduction to

GAS-LIQUID CHROMATOGRAPHY

An Introduction to
GAS-LIQUID CHROMATOGRAPHY

R. ALAN JONES

School of Chemical Sciences,
University of East Anglia, Norwich, England

1970

ACADEMIC PRESS · LONDON AND NEW YORK

ACADEMIC PRESS INC. (LONDON) LTD
Berkeley Square House
Berkeley Square
London W1X 6BA

U.S. Edition published by
ACADEMIC PRESS INC.
111 Fifth Avenue
New York, New York 10003

Copyright © 1970 by ACADEMIC PRESS INC. (LONDON) LTD

All Rights Reserved
No part of this book may be reproduced in any form by photostat, microfilm, or any other means, without written permission from the publishers

Library of Congress Catalog Card Number: 73-129789
ISBN: 0-12-389850-1

PRINTED IN GREAT BRITAIN BY
ADLARD & SON LTD
BARTHOLOMEW PRESS
DORKING

PREFACE

Progress in gas-liquid chromatography has been so rapid during the past decade that few organic chemists do not now use the technique routinely either for analysis or for the preparative separation of reaction products. Many books are available which describe the theory of chromatographic separation; there are, however, few sources outside the research literature to which the organic chemist can turn for general information on the practical aspects of the technique. The objective of this book is to provide a basic introduction to the practice of gas-liquid chromatography for organic chemists who have little knowledge of the technique.

In Chapter 1 fundamental theory is presented which allows the reader to understand the basic mechanism of the chromatographic process. The remaining Chapters describe the practical aspects of the operation of a gas-liquid chromatograph and the methods of handling and analysing micro amounts of organic compounds. Quantitative interpretation of the chromatogram and the general methods by which the components of a mixture may be analysed qualitatively are also presented. The book concludes with a reference section, set out in tabular form, which summarises the information considered to be of most use to the operator during routine GLC analysis. The reference section also includes a dictionary of gas-liquid chromatographic terms and abbreviations, a directory of manufacturers, and a brief survey of current commercially available analytical and preparative instruments.

During the preparation of this book I have become indebted to many of my colleagues for their helpful advice and constructive criticism, especially Dr. A. J. Boulton, Dr. D. G. Cooper, and Dr. A. McKillop. I am also grateful to Mr. T. Grey and Mr. D. Robinson of the Food Research Institute for many helpful discussions. The assistance afforded by many instrument manufacturers in the form of the generous provision of technical and scientific information is most gratefully acknowledged. The ultimate responsibility for the contents of the book, in particular any errors or omissions, is, however, entirely mine.

I should also like to thank Professor K. Hafner for his hospitality at the Institut für Organische Chemie, Darmstadt during the summer of

1969 when the greater proportion of the manuscript was prepared. Finally I thank my wife and family for their understanding and extreme patience during the past year.

Norwich R. ALAN JONES
June, 1970

ACKNOWLEDGMENTS

With the exception of those constructed by the author, the illustrations and tables in this book came from various sources and I am grateful to the following publishers and corporations for their permission to reproduce copyright material.

Academic Press, New York, N.Y.: Figs. 1, 4 and 25.
American Chemical Society, Washington, D.C.: Figs. 12, 20, 27, 30, 36, 53, 60, 63, 64, 65 and 67.
Applied Science Laboratories, Inc., Inglewood, California: Fig. 11.
Butterworth and Co. Ltd., London: Fig. 19.
Carlo Erba S.p.A., Milan, Italy: Fig. 31.
The Chemical Society, London: Figs. 54, 55 and 56.
Elsevier Publishing Co., Amsterdam, Holland: Fig. 35.
Fisher Scientific Company, Pittsburgh, Pa.: Fig. 13.
Hewlett Packard Ltd., Slough: Figs. 6, 70 and 71.
Interscience Publishers, Division of John Wiley and Sons, Inc., New York, N.Y.: Table IV.
The MacMillan Company, New York, N.Y.: Figs. 15, 16 and 17.
MacMillan (Journals) Ltd., London: Fig. 23.
National Research Council of Canada, Ottawa, Canada: Fig. 48.
Nester-Faust, Newark, Delaware: Fig. 74.
Perkin Elmer Ltd., Beaconsfield: Figs. 14, 73, 75 and 76.
Preston Technical Abstracts Co., Evanston, Illinois: Figs. 18, 49 and 65.
Pye-Unicam Ltd., Cambridge: Figs. 24, 28, 29, 77 (in part) and 78.
Smiths Industries Ltd., Wembley, London: Fig. 44.
Tracor Inc., Austin, Texas: Fig. 32.
Varian Associates Ltd., Walton-on-Thames: Fig. 9.
Wilmad Glass Co., Buena, New Jersey: Fig. 61.

CONTENTS

Preface v
Acknowledgments vii

Part I Theory and Practice

Chapter 1
Introduction

A. General Introduction and History 1
B. Basic Theory 4
   1. The Mechanism of Chromatographic Separation . . 4
   2. The Efficiency of Chromatographic Separation . . 7
   3. Retention Time and Retention Volume . . . 13
References 18

Chapter 2
The Basic Instrument for Gas-liquid Chromatography

A. Introduction 21
B. The Injection System 22
   1. Design of the Injection Port 22
   2. Introduction of the Sample 24
C. The Oven 30
D. The Column 31
   1. Packed Columns 31
   2. Capillary Columns 35
E. The Detector 37
   1. Introduction 37
   2. The Thermal Conductivity Detector or Katharometer . 42
   3. Ionisation Detectors 45
   4. Applications 52
F. The Flowmeter 53
G. The Carrier Gas 54
References 55

Chapter 3

The Principles of Operation

A. The Analysis of an Unknown Sample 57
   1. The Choice of a Suitable Column 57
   2. Selection of Gas Flow Rate and Column Temperature . 65
   3. Peak Attenuation 67
B. Pretreatment of the Sample 68
   1. Trimethylsilylation 71
   2. Esterification and Transesterification 73
   3. Acylation. 74
References 74

Chapter 4

Interpretation of the Chromatogram

A. Quantitative Analysis 77
   1. Introduction 77
   2. Measurement of Peak Areas 81
   3. Measurement of Peak Heights 87
   4. Non-Gaussian Peaks 88
B. Qualitative Analysis 91
   1. Use of Retention Data 91
   2. Functional Group Analysis 95
   3. Thin Layer Chromatography 96
   4. Physical Methods of Analysis 99
References 114

Chapter 5

Chromatographic Techniques

A. Temperature Programmed Gas Chromatography . . 117
   1. Introduction 117
   2. Variable Parameters 120
   3. Retention Temperature 122
B. Cryogenic Gas Chromatography 123
C. Flow Programmed Gas Chromatography 124
D. Backflushing 124

E.	Automated Preparative Chromatography.	127
	1. Preparative Columns	127
	2. Sample Introduction.	128
	3. Column Operational Cycle.	129
	4. Traps	130
F.	Automatic Routine Analysis and Product Control	130
References .		132

Part II   Appendix

A.	Glossary of Gas Chromatographic Terms, Symbols and Abbreviations .	137
B.	Operational Procedures and Common Operational Faults .	143
C.	Stationary Phases in General Use .	153
D.	Instrument Directory .	166
Company Directory .		188
Author Index .		193
Subject Index .		197

Part I
Theory and Practice

Chapter I

INTRODUCTION

A. General Introduction and History 1
B. Basic Theory 4
   1. The Mechanism of Chromatographic Separation . . . 4
   2. The Efficiency of Chromatographic Separation 7
   3. Retention Time and Retention Volume 13
References 18

A. General Introduction and History

The term *chromatography* includes all processes in which the separation of a mixture is accomplished by the differential adsorption or solution of the individual components of the mixture between two immiscible phases[1]. The common feature of all chromatographic separations is that one of the phases is *stationary* whilst the second phase is *mobile* and passes through the *stationary phase* "carrying" the individual components at different rates depending upon their relative distribution or *partition* between the two phases (see p. 4).

Gas-liquid chromatography (GLC) is a relatively recent development in the general field of chromatographic techniques but, in the last decade, it has become an increasingly popular method among organic chemists for analysis and for the isolation of individual components from complex mixtures. Compared with other chromatographic techniques, such as thin layer chromatography and column chromatography, the apparatus required for gas-liquid chromatography is relatively expensive and complicated and it owes its wide popularity primarily to its versatility and efficiency. It has, for example, the advantage of rapid simultaneous quantitative as well as qualitative analysis (see Chapter 4), and with very little modification of the basic apparatus, it is possible to efficiently separate the components of a mixture on a preparative scale (see Chapter 5). The stationary phase for gas-liquid chromatography is a liquid of low vapour pressure which is absorbed on a porous inert solid support enclosed in a narrow diameter tube. This arrangement is referred to as the chromatographic *column*. A closely related technique, *gas-solid chromatography*, employs a solid stationary phase. In both systems the *mobile phase* is an inert gas which is generally called the *carrier gas*. Both chromatographic processes are applicable only for the separation of gases or of liquids and solids which are readily vaporised below *c*. 300°.

A separation technique which may effectively be described as gas-solid chromatography was reported as early as 1931, but its potential as an analytical tool was not realised. The theoretical possibility of the use of the process for analysis was suggested in 1941 by Martin and Synge[2] as a further development of their work on liquid-liquid partition chromatography. The early studies of the new separation procedure were based on the use of a solid stationary phase, but the technique was found to be of little use as an analytical tool for the average organic chemist and it was soon displaced by gas-liquid chromatography. The rapid expansion in the use of the latter technique as an analytical tool can be attributed largely to the work of Martin and James who, in 1952, described the separation of a mixture of fatty acids by gas-liquid partition between nitrogen and a stationary phase of silicone oil containing 10% w/w stearic acid supported on kieselguhr[3, 4]. In the same year they also reported the separation of amines[5, 6]. In these first experiments the eluted compounds were detected and their concentration estimated by titration. The procedure was therefore limited in its use to the analysis of basic and acidic compounds. Also, the titration method was not sufficiently sensitive to measure very small concentrations of sample. However, the adaptation of the gas density balance for use as a detector, which was sensitive to small concentrations of compounds irrespective of their chemical structure, together with the development of the thermal conductivity detector gave a greater versatility to gas-liquid chromatography. With the manufacture of the first commercial instrument during 1955 improvements in design followed rapidly and a major advance in instrumentation came in 1957 with the ionisation detector (see p. 45) which had a considerably higher sensitivity and response than any other detector previously available. At about the same time Golay showed that a liquid phase coated on the interior surface of a capillary tube through which the flow of the carrier gas is unobstructed was more efficient for chromatographic separation than a conventional packed column. Extremely small quantities of sample must be used with capillary columns and it was fortuitous that their development coincided with the innovation of the ionisation detector. Further advances have produced detectors sensitive to specific elements and progress in the combination of gas-liquid chromatography with other physical analytical techniques and with computers has led to one of the most powerful analytical procedures.

1. INTRODUCTION

A measure of the success of gas-liquid chromatography as an analytical tool may be gauged by the hundredfold increase in the annual number of publications since 1952. By the end of 1968 well over 10,000 papers had been published describing not only the design or adaptation of gas chromatographic instruments but also reporting on almost every aspect of the application of gas-liquid chromatography to analytical problems.

As the number of publications increases, it is increasingly more important that the practical gas chromatographer should know the sources to which he can refer for information. The general abstracting services, such as *Chemical Abstracts*, provide a wide coverage of the current literature, but information is often more rapidly and conveniently obtained from specialist abstracting services.

Among the more useful sources of present day developments in gas-liquid chromatography are the various reports of the proceedings of symposia. The edited proceedings of the biannual symposia held in Europe under the auspices of the Gas Chromatography Discussion Group of the Institute of Petroleum are published regularly[7], and, at approximately three-monthly intervals, the same Discussion Group publishes abstracts of all publications which are of interest to gas chromatographers[8]. The proceedings of other symposia, which are held less frequently, are also published[9-13]. Current trends in gas chromatography are reviewed biannually in *Analytical Chemistry* and convenient sources of references to the current literature appear in the bibliographic section of the *Journal of Chromatography* and the *Journal of Chromatographic Science* (formerly the *Journal of Gas Chromatography*). Reviews of various aspects of gas chromatography are also to be found in *Advances in Chromatography*[14] and *Chromatographic Reviews*[15].

Although agreement in a uniform presentation of the retention data has still to be reached, compilations of data have been published by the American Society for Testing and Materials[16] and by the Preston Technical Abstracts Co.[17]. The latter company also provides a literature abstracting service presenting the abstracted information on punch cards for rapid sorting. They have also published a comprehensive index to the literature on gas chromatography up to 1962[18].

Several manufacturers publish periodicals which contain technical information of their instruments and reviews of the general applications of gas-liquid chromatography. These journals include *GC Newsletter* and *Instrument News* (Perkin-Elmer), *The Column* (Pye-Unicam), *Aerograph Research Notes* (Varian Aerograph), *Facts and Methods* (Hewlett-Packard), and *The Analyser* (Beckman). The addresses of these companies are to be found in Part II. With the proliferation of literature it is obvious that some uniformity in nomenclature should be established and a committee was set up in 1960 under the direction of the International Union of Pure and Applied Chemistry (IUPAC). The preliminary recommendations published by this committee[19] have been adhered to throughout this book.

B. Basic Theory

I. THE MECHANISM OF CHROMATOGRAPHIC SEPARATION

Let us first consider a single component sample. After introduction, the sample, if not already a gas, is vaporised and then swept by the carrier gas onto the column. On reaching the stationary phase the major part of the vapour is adsorbed, and an equilibrium is established between the adsorbed vapour and the smaller proportion remaining in the gas phase. This small proportion of unadsorbed vapour moves forward with the carrier gas where it again establishes an equilibrium condition with the stationary phase. Simultaneously, pure carrier gas reaches that point in the stationary phase where the originally adsorbed vapour was at equilibrium and causes a proportion of the vapour to re-enter the gas phase to restore the disturbed equilibrium condition. Thus, vapour is continuously leaving the stationary phase and entering the gaseous phase at the rear of the vapour zone and is being removed from the gas phase at the front of the zone. This movement of the vapour to and from the carrier gas is the fundamental mechanism of gas chromatography and results in the compact movement of the sample through the stationary phase. An ideal distribution of the sample between the stationary phase and the carrier gas is shown in Fig. 1. (As will be seen later, however, this ideal dynamic equilibrium is never attained in practice).

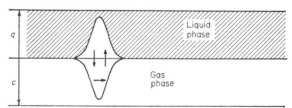

Fig. 1. Schematic diagram of an ideal column in which the sample is distributed in dynamic equilibrium between the gas and liquid phases.

The distribution of the vapour between the liquid phase (the stationary phase) and the gas phase is described by Henry's law:

$$q = \beta c$$

where q is the concentration of the vapour in the liquid phase and c is the concentration of the vapour in the gas phase. The constant β is a measure of the distribution of the vapour between the two phases and is conventionally called the *partition coefficient*. The value of this constant determines the rate at which the zone of vapour passes through the stationary phase. Thus, compounds which are preferentially adsorbed by the stationary phase and which consequently move slowly through the column have a larger partition coefficient than those compounds which are not readily adsorbed.

In discussing the value of the partition coefficient it is necessary to define the form in which the concentrations of the vapour in the two phases are described. A dimensionless value of the partition coefficient is obtained only when the concentrations of the vapour in both the liquid and the gas phase are given in the same units, as for example, grams per cc, and also at the same temperature, which is usually taken as that of the stationary phase (the column temperature). This fundamental form of Henry's law

$$\alpha = \frac{\text{Weight of vapour per cc of stationary phase}}{\text{Weight of vapour per cc of gas at column temperature}}$$

is difficult to utilise as the volume of the stationary phase is not readily ascertainable.

It is considerably easier to determine the weight of the stationary phase and the more usual form of Henry's law is given as:

$$\beta = \frac{\text{Weight of vapour per gram of stationary phase}}{\text{Weight of vapour per cc of gas at column temperature}}$$

Thus, the two forms of Henry's law are related by the expression

$$\alpha = \beta\rho$$

where ρ is the density of the stationary phase at the temperature at which α and β are determined.

A third form of the equation, which should be used in any calculation of thermodynamic properties of the vapour from the chromatographic retention data, but which is not normally considered in the analytical applications of GLC, utilises the weight of the vapour per cc of the gas phase corrected to 0°C (273°K) as a reference temperature.

$$\gamma = \frac{\text{Weight of vapour per gram of stationary phase}}{\text{Weight of vapour per cc of gas phase at 273°K}}$$

This form of the partition coefficient is therefore related to β by the equation.

$$\gamma = \beta \frac{273}{T_c}$$

where T_c (°K) is the column temperature.

Although the sample may be considered to be introduced onto the stationary phase over an infinitely short interval of time so that the vapour initially occupies an extremely narrow zone, the inability of the chromatographic system to establish an ideal dynamic equilibrium distribution of the vapour between the stationary and mobile phases results in a broadening of the input distribution. A column which functions in this manner is said to be *non-ideal*, whereas the hypothetical column which does not broaden the input distribution is called an *ideal* column. The shape of the output distribution curve of the eluted sample depends to a large extent upon the manner in which the sample vapour interacts with the stationary phase. At any constant temperature, if the distribution of the vapour between the gas and liquid phases obeys Henry's law then the equation describes a linear *isotherm* of interaction between the phases (Fig. 2a) and the resultant broadening of the input distribution gives a symmetrical Gaussian output distribution. If the vapour, instead of dissolving in the stationary phase, is only adsorbed, then Henry's law is not obeyed over the full range of concentration. The concentration of the vapour in the stationary phase becomes less than that required for simple proportionality as the concentration in the gas phase increases. This gives the effective result that the partition coefficient is lower at higher concentrations, and hence the middle of the vapour zone tends to travel faster than the edges. The assymetry of the

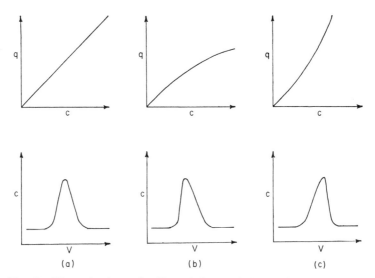

Fig. 2a. Linear isotherm for Henry's law and the resultant symmetrical band. b. and c. Skewed bands resulting from negative and positive deviations from Henry's law.

eluted zone depends upon the degree of curvature of the isotherm (Fig. 2b). Conversely, if the concentration of the vapour in the stationary phase increases proportionately more rapidly than the increase in its concentration in the gas phase as the concentration increases, the centre of the vapour zone moves at a slower rate than the leading edge (Fig. 2c). This second type of non-linearity is rare.

Asymmetry of the eluted band may be conveniently avoided by the use of extremely low concentrations. Under these conditions the isotherms are almost linear irrespective of the solute : solvent system.

2. THE EFFICIENCY OF CHROMATOGRAPHIC SEPARATION

It is self-evident from the previous section that if a mixture of two or more components having widely different partition coefficients is introduced onto a column then they travel at different rates and are eluted from the column separately. Although the separation of the components is related primarily to the ratio of their partition coefficients, it is also dependent upon the broadness of the eluted bands. The efficiency of the separation may be defined therefore in terms of the ability of the column to separate a given sample into its component compounds such that they are eluted as narrow bands which do not overlap.

An analogy, using the "theoretical plate" concept, can be made between the efficiency of the chromatographic column and that of a fractional distillation column. The idea of a "theoretical plate" is best appreciated, however, by reference to countercurrent liquid-liquid extractions where the separation of the components is effected in a series of discrete equilibria. Each equilibrium step is called a "theoretical plate". For gas chromatographic separations the "theoretical plate" is defined as the length of column required to establish an equilibrium in which the vapour pressure of the solute in the gas phase equals the vapour pressure of the solute in the liquid phase. The performance of the column may then be described in terms of the *plate height*, h, or the *height equivalent to a theoretical plate* (HETP). The smaller the value of the plate height the more efficient is the column. As the gas phase is continuously moving however, the equilibrium is never attained and it is more realistic to define the column performance in terms of the band broadening per unit length of column. This may be expressed as

$$h = \sigma^2/l$$

where the band broadening is described in terms of the square of the standard deviation of the band, σ, and l is the length of the column.

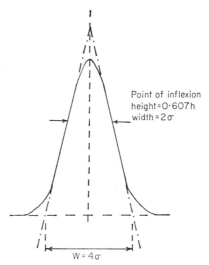

Fig. 3. Relationship between standard deviation and the width of a Gaussian peak.

Since the relative standard deviation, i.e. the band broadening per unit length of column, σ/l, may be described as a function of the retention of the solute on the column as follows

$$\frac{\sigma}{l} = \frac{W}{4V_{R'}}$$

where W is the base width of the band (see Fig. 3) and $V_{R'}$ is the retention volume of the solute (see p. 13) it is possible to express the plate height in terms of readily measured parameters:

$$h = \frac{l}{16(V_{R'}/W)^2}$$

The relative performance of different columns of the same length can therefore be obtained from the value of the denominator, $16(V_{R'}/W)^2$, and is called the theoretical *plate number* of the column. The higher the value of the plate number, the more efficient is the column.

Several factors contribute to the broadening of the band on the column. Of considerable importance is the velocity of the carrier gas. This is related to the plate height by the *van Deemter equation*:

$$h = A + \frac{B}{v} + C \cdot v$$

where v is the velocity of the carrier gas. The first term, A, is the so-called "eddy diffusion term", which is zero for capillary columns (see p. 36) and small for packed columns when the stationary phase consists of uniformly small particles. The term B/v is a band broadening factor due to longitudinal diffusion of the sample vapour in the gas phase. This diffusion occurs whether the vapour zone is stationary or is being eluted. Thus, the longer the sample is retained on the column then the broader is the eluted band. It follows therefore that at a constant carrier gas flow-rate the use of long columns produces a greater band broadening due to longitudinal diffusion than does a short column.

In practice the carrier gas is maintained at a high velocity in order to reduce analysis time and consequently the last term of the van Deemter equation is the most important factor. It relates to the non-instantaneous equilibration of the sample between the gas and the liquid phases. Instead of the ideal situation in which the vapour zone in the gas phase is always at equilibrium with the solute in the liquid phase, the solute is never able to attain an equilibrium state before the vapour zone has been swept forward by the carrier gas. The constant C comprises the gas phase mass transfer term, C_g, and the liquid phase mass transfer term, C_l. The value of C_g is proportional to the square of the diameter of the stationary phase particles. The use of smaller particles in packed columns should therefore result in a higher column efficiency, but owing to the difficulty in packing such columns uniformly, the eddy diffusion term increases with a resultant lowering of the column efficiency. The optimum particle size appears to be within the range 50 to 80 B.S.S. although, for most practical purposes, somewhat higher particle sizes are preferred. The C_l term is dependent upon the ratio of the liquid phase to solid support, i.e. the effective thickness of the liquid phase. It is also inversely proportional to the diffusion coefficient of the solute in the liquid phase and, as the diffusion coefficient varies inversely with viscosity, the column efficiency is improved by the use of low viscosity liquid phases.

The value of the partition ratio of the solute between the gas and liquid phases varies inversely with temperature and, as both C_g and C_l are functions of the partition coefficient, it is these terms which are principally responsible for the change in the column efficiency with a change in temperature. There will also be a small change in the value of the term B as a consequence of an increase in the coefficient of diffusion of the solute in the gas phase with an increase in temperature. The overall effect of temperature upon column efficiency is extremely complex but, in general, the efficiency is increased by lowering the temperature of the column. However, as a result of the lower temperature, the retention of the solute on the column increases and thereby gives a greater opportunity for the band to broaden.

The temperature of the sample injector is also important for high column efficiency. To obtain a narrow input distribution of sample onto the column, it is necessary to have the injection temperature higher than the boiling point of the least volatile component of the sample, thus ensuring rapid and complete vaporisation. As the temperature of the injector block is further increased, the column performance improves until a temperature is reached at which the breadth of the input distribution is small compared with the broadening effect of the column. Beyond this critical temperature no further improvement in column efficiency can be attained by altering the injection temperature.

The introduction of large samples at low injection temperatures results in considerable band broadening since complete vaporisation of the sample takes longer and the total heat of vaporisation is high. Hence, the amount of sample which can be efficiently separated is limited. The sample capacity of a column is usually defined as that amount which can be applied without more than 10% loss in column efficiency and is directly proportional to the amount of stationary phase per unit length of column. It is consequently proportional to the square of the column diameter. The effect of overloading the column is observed as a loss of resolution with band broadening and an asymmetry in the band shape with a resultant apparent change in the retention of the solute by the column. This loss in column performance originates in two factors. Firstly, because of the finite volume of the vapour injected the input distribution even at high temperatures is broadened by large samples, and secondly, it is more difficult for the vapour zone of a large sample to attain a dynamic equilibrium between the gas and stationary phases. If we assume that the sample is completely vaporised, i.e. the column temperature is above the boiling point of the sample, then the excess of sample beyond its solubility limit in the stationary phase moves at the same velocity as the carrier gas. This produces a band with a sharp profile and a diffuse tail (Fig. 4a). The converse effect is observed if the column temperature is below the boiling point of the sample. The excess of sample remains in a condensed state and its velocity through the column depends entirely upon its solubility in the stationary phase. The bulk of the sample therefore travels more slowly than the carrier gas and the eluted band has a diffuse profile and a sharp tail (Fig. 4b). In both situations the observed position of the band maximum differs from the true value as measured for an infinitely small sample. As is to be expected, the column temperature has a considerable effect upon the band shapes, particularly upon the type shown in Fig. 4b. Not only does the band width decrease with an increase in temperature but there is also a decrease in the asymmetry of the band as a result of the higher solubility of the solute in the stationary phase at the higher temperature.

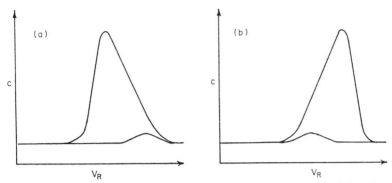

FIG. 4. Peak shapes and retention volumes for very large and infinitely small sample volumes in nonlinear chromatography.

So far in this discussion of column efficiency, only the band broadening of a single component system has been considered. It is obviously important to relate column efficiency with the band widths and retention of two adjacent bands. The *resolution*, R, of two adjacent bands of equal area can be defined as the ratio of the separation of the band maxima, ΔV, and their average base width.

$$R = \frac{2\Delta V}{w_1 + w_2}$$

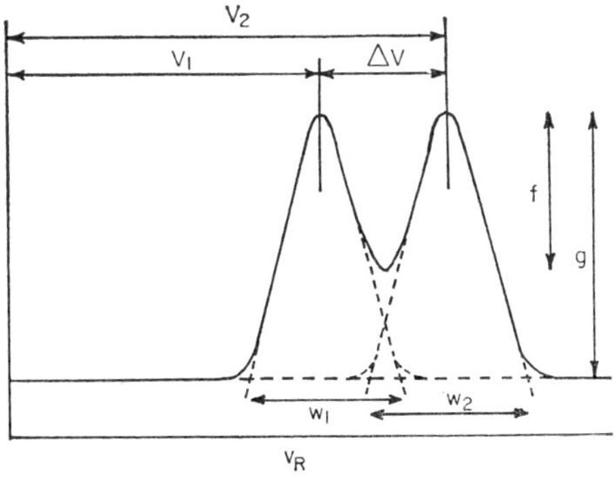

FIG. 5. Separation of adjacent peaks.

In general the bands are considered to be completely resolved when $R = 1\cdot 5$. At this degree of resolution there is only 0·3% overlap of the two bands. However, for most practical purposes, a separation of 98% is acceptable, when the resolution is equal to unity. Kaiser[20] defines resolution in terms of the ratio of the valley depth between the overlapping bands and the average band height, $R = f/g$. As this convention is contrary to that recommended by the IUPAC committee[19] and is not widely used, it is not considered further. As the resolution is inversely proportional to the widths of the bands, it is obvious that the factors which control column performance are important in the determination of the resolution attainable with any particular column. The resolution may be improved by increasing the separation of the band maxima either by changing the stationary phase or by operating the column at a lower temperature. In practice one is forced to make a compromise in the choice of variables to obtain optimum resolution with a minimum loss in column performance. Thus for example, although a low column temperature may give the best separation of the band maxima, the high probability of an increase in the widths of the bands can result in a poorer resolution.

As the band width can be described in terms of column efficiency using the plate height (see p. 8), the resolution of two bands having identical base widths is given by the expression:

$$R = \frac{\Delta V}{4(1+V_2)} \cdot \frac{l}{h}$$

where V_2 is the retention volume (see p. 13) of the slowest band and ΔV is the separation of the maxima. Thus, if the plate height is known, it is possible to calculate the minimum column length which will effect a 98% separation of two adjacent bands. For complete separation of the bands, $R = 1 \cdot 5$, the factor 16 should be replaced by 36.

$$l_{\min} = 16h(1+V_2)/\Delta V^2$$

The above discussion assumes that the band distribution is Gaussian and implies that the two components do not interact with each other when they overlap on the column. However, compounds which are structurally similar can change each other's chromatographic behaviour, each serving as a stationary phase for the other in the overlap region. This effectively increases their solubility in the stationary phase in that region and produces an asymmetry in the band profiles (Fig. 6) with a resultant loss of resolution. Such a situation arises in particular when one of the components is in excess such that it overloads the column. Under these conditions not only is the band for the overloaded component asymmetric but so is that for the non-overloaded component. The degree of overlap of the two bands therefore depends upon the order of their elution (Fig. 6). If the minor constituent of the mixture shown in Fig. 6 is to be collected, a stationary phase should be chosen such that it is eluted after the major component. When both components are present in similar concentrations, it is always easier to collect a purer sample and larger volume of the second component.

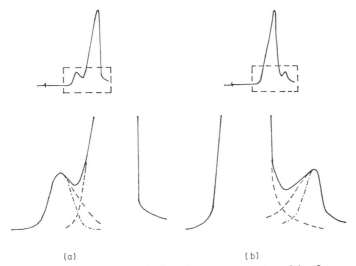

FIG. 6. Asymmetry in the peak of a minor component resulting from overlap with a component of extremely high concentration. a. Minor component preceding major component. b. Minor component following major component. ·········· Gaussian peak - - - - - - Actual peak shape.

3. RETENTION TIME AND RETENTION VOLUME

The interval of time between the introduction of the mixture onto the column and the elution of a component is referred to as the *retention time* for that particular component. The value of the retention time, as well as being directly related to the partition coefficient, is a function of several variables of operation. An obviously important factor is the flow rate of the carrier gas. Assuming no change in the carrier gas pressure along the length of the column, the retention time is inversely proportional to the flow rate of the gas. The product of the flow rate and the retention time is equivalent to the volume of carrier gas required to carry a component through the stationary phase and is characteristic of an individual compound. This volume is called the *retention volume* and is independent of the flow rate of the carrier gas. It is, however, dependent upon other variables and, therefore, in comparing the retention volumes of different compounds the operating conditions should be comparable. For the ideal situation in which the input distribution of the sample is infinitely small and the band broadening is negligible the retention volume can be shown to be related to the partition coefficient β, by the equation:

$$V_R = l(a+m\beta)$$

where l is the length of the column, a is the volume of the gas phase per unit length of column, and m is the mass of the stationary phase per unit length of column.

The product $a \cdot l$ is the total volume of the carrier gas in the column and is equivalent to the retention volume of a non-adsorbed sample. This volume comprises the interstitial volume of the stationary phase and also the effective volumes of the sample injector and the detector. This volume is given the symbol V_M and is called the *total gas hold-up* or the *dead volume* of the apparatus. The product $m \cdot l$ is the total weight of the stationary phase, W. Thus, the retention volume comprises two parts, the second of which, from a practical point of view, controls the variation in the observed retention volumes.

$$V_R = V_M + W\beta$$

Measurement of the retention volume of a non-adsorbed substance, such as air or a permanent gas other than the carrier gas, gives directly a value for the dead volume.

$$\beta = 0$$
$$\therefore V_R = V_M$$

Correction of the observed retention volume for the dead volume of the chromatographic apparatus gives the *adjusted retention volume*, $V_{R'}$,

$$V_{R'} = V_R - V_M$$

It is not always possible to measure directly the retention volume of the non-adsorbed compound. For example, flame ionisation detectors are insensitive to permanent gases. An alternative method by which the dead volume may be measured uses the correlation between the adjusted retention volume and the carbon number of homologous organic compounds whereby the retention volumes of successive homologues follow a geometrical progression. This is illustrated in Fig. 7a in which x/y equals y/z. It follows that there is a linear relationship between the logarithm of the adjusted retention volumes and the carbon numbers (Fig. 7b).

A more systematic method involves the measurement of the retention volumes of three homologous compounds, the carbon numbers of which satisfy the relation $C_m - C_l = C_n - C_m$. If y_l, y_m and y_n are the distances of their peaks from an arbitrary point, then the distance x of that point from the retention volume corresponding to the dead volume is given by the equation:

$$x = \frac{y_l^2 - y_l y_n}{y_n + y_l - 2y_m}$$

So far we have assumed that the carrier gas is maintained at a constant pressure whilst passing through the stationary phase. This assumption, however, is invalid. The viscosity of the carrier gas and its restricted

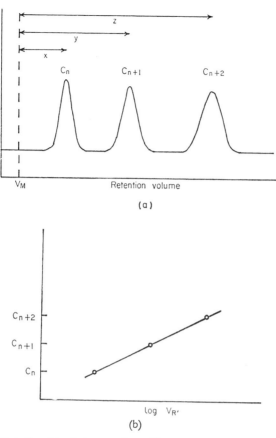

Fig. 7a. Relationship between the adjusted retention volumes, x, y and z, and the carbon number of homologous organic compounds. b. Linear relationship between the logarithm of the adjusted retention volumes and the carbon number of homologous organic compounds.

passage through the interstitial spaces of the stationary phase produces a pressure gradient along the length of the column. There is, therefore, a higher pressure of gas at the inlet than at the outlet of the column and since the molecular rate of flow of the gas is constant throughout the column, the volume flow rate of the gas is greater at the outlet than at the inlet of the column. As might be expected, the pressure gradient, and the resultant velocity gradient, affect the retention volume. The *corrected retention volume*, V_{R^0}, which takes into consideration the pressure gradient, is given by the equation:

$$V_{R^0} = jV_R$$

where
$$j = \frac{3(P_i/P_o)^2 - 1}{2(P_i/P_o)^3 - 1}$$

P_i and P_o being the pressures of the carrier gas at the inlet and outlet of the column respectively. Correction of the adjusted retention volume for the pressure gradient gives the *net retention volume*, V_N.

$$V_N = jV_{R'}$$
$$= jV_R - V_M$$

The correction for the pressure gradient is already incorporated in the value of the dead volume, V_M.

The pressure gradient is usually calculated from the measured flow-rates of the carrier gas at the inlet and outlet of the column and there is invariably a difference between the actual flowrate in the column and the measured flowrates due to the difference in temperatures of the column and the flowmeters. By Charles' law the variation in the volume, V, of a gas resulting from a change in the temperature, T, is given by:

$$V_c = V_f \frac{T_c}{T_f}$$

where the subscripts c and f refer to the column and flow-meter. Hence the corrected flow rate of the carrier gas through the column is obtained by multiplying the measured flow rate by T_c/T_f. More accurately T_f is the temperature at which the flowmeter is calibrated.

Retention volumes are dependent upon the column temperature, not only because of the change in the flowrate of the carrier gas with the change in column temperature, but also as a result of a change in the partition coefficients of the adsorbed compounds. For a system operating at a constant carrier gas flowrate, the variation in the value of the partition coefficient, γ, with a change in the column temperature is dependent upon the enthalpy of evaporation of the solute from the liquid to the gas phase, ΔH_v.

$$\frac{d \ln \gamma}{dT} = \frac{-\Delta H_v}{RT^2}$$

The partition coefficient, γ, is equivalent to the *specific retention volume* of the solute, V_g. Hence, the logarithm of the retention volume is inversely proportional to the column temperature.

$$\ln V_g = \frac{-\Delta H_v}{RT}$$

In the comparison of the retention volume data of identical compounds measured on different instruments but using the same type of stationary phase, the preceding discussion shows that it would be necessary to correct for (a) the differing dead volumes of the two instruments, (b) the differing pressure gradients, (c) any variation in the operating temperature of the column, and (d) the total weights of the stationary phase in the columns of each instrument. Correction (a) is an additive term, whereas for (b), (c) and (d) the observed retention volume has to be multiplied by the correction factors. The variable operational parameters affect the retention volumes of all compounds measured simultaneously by similar factors. Thus, for routine analysis one can eliminate the work involved in the correction of these variables by using an internal reference compound. The ratio of the adjusted retention volume of the compound under investigation to that of the reference compound is called the *relative retention*.

$$r = \frac{V_{R'}}{V_{R'\text{ref}}}$$

Alternatively, as the flow rate of the carrier gas can be assumed to remain constant during the chromatographic measurements, direct comparison of the corrected retention times can be used to obtain the relative retention. The two calculations are equivalent and correspond directly to the ratio of the partition coefficients of the compounds under investigation and the reference compound.

Partition coefficients depend upon the intermolecular interactions between the adsorbed solutes and the liquid phase and are determined not only by the chemical structure of the solutes but also by the chemical nature of the liquid phase (see p. 58). It is not possible, therefore, without a considerable knowledge of the thermodynamics of the systems to make a direct comparison of relative retention data, obtained using the same reference compound, from two different types of stationary phase. For, although the partition coefficient of, for example, the reference compound may remain relatively unchanged there could be a considerable difference in the partition coefficients for the other compounds on changing the stationary phase.

In the determination of the relative retention of a series of compounds the chosen reference compound should have a retention volume which falls near the middle of those for the series of compounds. The IUPAC committee[19] has recommended that wherever possible one of the following standards should be used:

n-butane	naphthalene
2,2,4-trimethylpentane (*iso*-octane)	ethyl methyl ketone
benzene	*cyclo*hexanone
p-xylene	*cyclo*hexanol

These compounds cover a wide range of retention volumes and are suitable for medium temperature analysis up to 150°C. Above this temperature it is recommended that some other readily available commercial chemical be used. Other standards should be used if the need arises, as, for example, if the retention volume of the most suitable standard compound chosen from the above list overlaps or coincides with that of one of the components of the mixture being analysed.

This book is intended primarily as an introduction to the practice of gas-liquid chromatography. For a more detailed discussion of the principles of the technique the reader is directed to the other texts which are available[21-25]. Extended texts describing the practical aspects[26] and applications of gas-liquid chromatography[27] have also been published.

References

1. For a general survey of chromatography see Cassidy, H. G., "Fundamentals of Chromatography", Vol. X of "Technique of Organic Chemistry" (A. Weissburger, ed.), Wiley, New York (1957).
2. Martin, A. J. P. and Synge, R. L. M., *Biochem. J.*, **35**, 1358 (1941).
3. James, A. T. and Martin, A. J. P., *Analyst*, **77**, 915 (1952).
4. James, A. T. and Martin, A. J. P., *Biochem. J.*, **50**, 679 (1952).
5. James, A. T., Martin, A. J. P. and Smith, G. H., *Biochem. J.*, **52**, 238 (1952).
6. James, A. T., *Biochem. J.*, **52**, 242 (1952).
7. Symposia of the Gas Chromatography Discussion Group of the Institute of Petroleum, London. "Vapour Phase Chromatography", Proceedings of the First Symposium, London, 1956 (D. H. Desty, ed.) Academic Press, New York (1957). "Gas Chromatography, 1958", Proceedings of the Second Symposium, Amsterdam, 1958 (D. H. Desty, ed.) Academic Press, New York (1958). "Gas Chromatography, 1960", Proceedings of the Third Symposium, Edinburgh, 1960 (R. P. W. Scott, ed.) Butterworths, London (1960). "Gas Chromatography, 1962", Proceedings of the Fourth Symposium, Hamburg, 1962 (M. van Swaay, ed.) Butterworths, London (1962). "Gas Chromatography 1964", Proceedings of the Fifth Symposium, Brighton, 1964 (A. Goldup, ed.) Institute of Petroleum, London (1964). "Gas Chromatography, 1966", Proceedings of the Sixth Symposium, Rome, 1966 (A. B. Littlewood, ed.) Institute of Petroleum, London (1967). "Gas Chromatography, 1968", Proceedings of the Seventh International Symposium, Copenhagen, 1968, (C. L. A. Harbourn and R. Stock, eds), Institute of Petroleum, London (1969).
8. "Gas Chromatography Abstracts", (C. E. M. Knapman, ed.) published annually by Butterworths, London (1958-1962) and Institute of Petroleum, London (1963 onwards).
9. Symposia of the Instrument Society of America. "Gas Chromatography", Proceedings of the First Symposium, 1947, (V. J. Coates, H. J. Noebels and I. S. Fagerson, eds) Academic Press, New York (1958). "Gas Chromatography", Proceedings of the Second Symposium, 1959, (H. J. Noebels, R. F. Wall and N. Brenner, eds) Academic Press, New York (1961). "Gas Chromatography", Proceedings of the Third Symposium, 1961. (N. Brenner, J. E. Callen and M. D. Weiss, eds) Academic Press, New York (1962). "Gas

Chromatography", Proceedings of the Fourth Symposium, 1963, (L. Fowler, ed.) Academic Press, New York (1963).
10. Proceedings of Symposia on Advances in Gas Chromatography, Houston, Texas, *Anal. Chem.*, **35**, 426 (1963); *Anal. Chem.*, **36**, 1410 (1964) and "Advances in Gas Chromatography—1965", (A. Zlatkis and L. Ettre, eds) Preston Technical Abstracts Co., Evanston, Ill. (1966).
11. Proceedings of the Cansisius College Gas Chromatography Institute, Buffalo, New York. "Progress in Industrial Gas Chromatography", (H. A. Szymanski, ed.) Plenum Press, New York (1961). "Lectures on Gas Chromatography", (H. A. Szymanski, ed.) Plenum Press, New York (1963). "Biomedical Applications of Gas Chromatography", (H. A. Szymanski, ed.) Plenum Press, New York (1964).
12. Symposia of the East German Gas Chromatography Group. "Gas Chromatographie, 1958", Proceedings of the First Symposium, Leipzig, 1958 (H. P. Angele, ed.) Akademie-Verlag, Berlin (1959). "Gas Chromatographie, 1959", Proceedings of the Second Symposium, Böhlen, 1959 (R. E. Kaiser and H. G. Struppe, eds) Akademie-Verlag, Berlin (1959). "Gas Chromatographie, 1961", Proceedings of the Third Symposium, Schkopau, 1961 (H. P. Angele and H. G. Struppe, eds) Akademie-Verlag, Berlin (1961). "Gas Chromatographie, 1963", Proceedings of the Fourth Symposium, Leuna, 1963 (H. P. Angele and H. G. Struppe, eds) Akademie-Verlag, Berlin (1963). "Gas Chromatographie, 1965", Proceedings of the Fifth Symposium, East Berlin, 1965 (H. G. Struppe, ed.) Akademie-Verlag, Berlin (1965).
13. Symposia of the GAMS. "Séparation Immédiate et Chromatographie, 1961", (J. Tranchant, ed.) GAMS, Paris (1961). "Atti delle Giornate Italiane della Separatione Immediate e della Chromatographia", Societa Italiana per lo Studio della Sostanze Grasse, Milan (1964). "Chromatographie et Méthodes de Séparation Immédiate", (G. Parissakis, ed.) Union of Greek Chemists, Athens (1966).
14. "Advances in Chromatography", 1965 onwards (J. C. Giddings and R. A. Keller eds) Edward Arnold, London.
15. "Chromatographic Reviews", 1959 onwards (M. Lederer, ed.) Elsevier, Amsterdam.
16. Lewis, J. S., "Compilation of Gas Chromatographic Data", American Society for Testing and Materials, Philadelphia (1963).
17. McReynolds, W. O., "Gas Chromatograph Retention Data", Preston Technical Abstracts Co., Evanston (1966).
18. Preston, S. T. and Hyder, G., "A Comprehensive Bibliography and Index to the Literature on Gas Chromatography", Preston Technical Abstracts Co., Evanston (1965).
19. "Preliminary Recommendations on Nomenclature and Presentation of Data in Gas Chromatography", *Pure and Appl. Chem.*, **1**, 177 (1960).
20. Kaiser, R., "Gas Phase Chromatography" Vol. 1, p. 39, Butterworths, London (1963).
21. Littlewood, A. B., "Gas Chromatography, Principles, Techniques, and Application," Academic Press, New York (1962).
22. Purnell, J. H., "Gas Chromatography", Wiley, New York (1962).
23. Dal Nogare, S. and Juvet, R. S., "Gas-Liquid Chromatography. Theory and Practice", Interscience, New York (1962).
24. Keulemans, A. I. M., "Gas Chromatography", Reinhold, New York (1959).

25. Schupp, O. E., "Gas Chromatography", Vol. XIII of "Technique of Organic Chemistry", (E. S. Perry and A. Weissberger, eds) Wiley, New York (1968).
26. Ettre, L. S. and Zlatkis, A., "The Practice of Gas Chromatography", Interscience, New York (1967).
27. Burchfield, H. P. and Storrs, E. E., "Biochemical Applications of Gas Chromatography", Academic Press, New York (1962).

Chapter 2

THE BASIC INSTRUMENT FOR GAS-LIQUID CHROMATOGRAPHY

A. Introduction 21
B. The Injection System 22
   1. Design of the Injection Port 22
   2. Introduction of the Sample 24
C. The Oven 30
D. The Column 31
   1. Packed Columns 31
   2. Capillary Columns 35
E. The Detector 37
   1. Introduction 37
   2. The Thermal Conductivity Detector or Katharometer . . 42
   3. Ionisation Detectors 45
   4. Applications 52
F. The Flowmeter 53
G. The Carrier Gas 54
References 55

A. Introduction

A typical arrangement for gas-liquid chromatography is shown schematically in Fig. 8. The flow of the carrier gas from the pressurised cylinder is controlled by reduction valves and monitored by a pressure gauge and flowmeter. The gas passes through the sample injection block, S, into which the sample to be analysed can be introduced. The sample is carried by the carrier gas onto the column from which the individual components are progressively eluted. The carrier gas and the eluted components pass through the detector, D, which causes the passage of the component to be recorded on the recorder, and they finally emerge from the apparatus via a second flowmeter. If the eluted sample is to be collected, a stream splitter is fitted between the column outlet and the detector in order that a major proportion of the eluted sample by-passes the detector and is trapped in the collector unit, C. The use of two flowmeters, as described here, aids the measurement of the pressure gradient along the column.

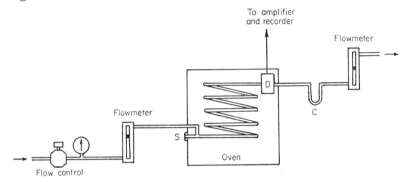

FIG. 8. Schematic diagram of a typical gas chromatograph.

B. The Injection System

1. DESIGN OF THE INJECTION PORT

The design of the inlet system varies with the manufacture of the chromatograph. Figure 9 illustrates a typical injection block of simple design. The injection port is sealed with a rubber septum through which the needle of the syringe holding the sample is inserted directly into a tube attached to the chromatographic column. The temperature of the injection block is controllable and is usually adjusted so as to be approximately 50° higher than that of the column. The high temperature ensures flash evaporation of the sample and the carrier gas sweeps the vaporised sample onto the stationary phase. Flash evaporation is necessary if the input distribution of the sample on the column is to be narrow. This is particularly important where the volume of the sample is large as, for example, with preparative gas chromatography (see p. 127). For accurate measurement of retention volumes, flash evaporation of all

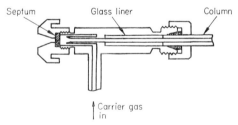

FIG. 9. A simple injection block.

the components of the sample is imperative, otherwise the more volatile components will be swept onto the column before those components having higher boiling points. The temperature of the injection block should therefore be higher than the boiling point of the least volatile component. Also, to prevent the possibility of condensation of the sample prior to it reaching the stationary phase and to minimise any effects of the small decrease in the temperature of the carrier gas as a result of the heat absorbed by the sample on vaporisation, the carrier gas is usually preheated, as illustrated in Fig. 10, before it enters the injection block.

As indicated in Chapter 1, the efficiency of chromatographic separation depends, amongst other factors, upon the size of the sample. The critical sample size for ideal operation is governed by the number and relative concentrations of the components of the sample and also by the type of

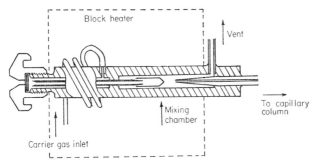

FIG. 10. A heated injection block with stream splitter

column. For standard packed columns, the usual sample size is of the order of 1 μl, however, the sample capacity of narrow diameter or capillary columns (see p. 35) is of the order of 10^{-3} μl. Samples of this volume may be introduced by means of a stream splitter. Approximately 1 μl of sample is injected into the injection block in the normal manner and is allowed to vaporise. The homogeneous mixture of the sample and the carrier gas is then divided into unequal parts, the smaller of which is introduced onto the chromatographic column whilst the remainder is discarded. Basically, the dynamic stream splitting system consists of two concentric tubes (Fig. 10). To obtain a completely homogeneous mixture of the sample and the carrier gas, a turbulent flow is required and this is attained by a specially designed mixing chamber. To prevent condensation of the sample, the carrier gas is preheated and the entire injection block and stream splitter is enclosed in a sheath at a constant high temperature. Back diffusion of the discarded vapours is prevented by the incorporation of a capillary orifice in the vent. This also acts as a means of controlling an accurate and constant split ratio.

Thermally sensitive compounds which are unstable when in contact with metal surfaces due to a catalytic effect of the metal cannot be introduced satisfactorily onto the chromatographic column using the injection block described above. Instead, a direct on-column injection procedure is followed using an injection port similar to that shown in Fig. 9 but modified so that the central concentric tube is part of the column. Contact between the sample and the metal surfaces of the injection block is eliminated by the use of a glass liner and the sample is delivered from the syringe needle within a few centimetres of the stationary phase so that it has minimum contact time with the injection block. In using on-column injection it is also advisable to use a glass column and a stationary phase which has no catalytic effect upon the stability of the sample. The catalytic effect of metallic injection ports and stream splitters may also be eliminated by coating their interior surfaces with an involatile silicone gum, such as SE-30 or OV-1. The surfaces are conveniently coated by injection of the gum from a syringe into the heated injection block under a slow carrier gas flow. (The column is not attached during the conditioning.)

2. INTRODUCTION OF THE SAMPLE

The sample must be gaseous or an easily vaporised liquid or solid. The sample size for analytical work will depend upon the dimensions of the column (see p. 10) and upon the sensitivity of the detector (see p. 53) but, for convenience of introduction into the chromatograph, the sample size usually lies in the range of 0·5 to 2·0 μl. The method of injection differs according to the physical state of the sample.

(a) *Liquid Samples:* With the exception of gases or extremely volatile liquids, most organic compounds may be introduced onto the chromatographic column in the form of a liquid sample, either as the neat compound or in the case of solids (and also liquids) as a solution. It is most convenient to inject the sample through the rubber septum in the injection port by means of a syringe. Precision syringes are available with capacities ranging from 0·5 to 10 μl and larger, or alternatively a micrometer syringe can be used. However, even with extreme care, injection of an absolute amount of material is difficult and may not be reproducible. This is because the sample must be injected into a system having a pressure greater than atmospheric and also having a high temperature. The needle of a 10 μl syringe holds approximately 1 μl of sample and, depending upon the length of time the operator holds the syringe in the injection port, this volume of sample may vaporise and enter the chromatographic system. Thus, if the period of injection time is long, the sample zone introduced onto the column is broadened and the resolution of the components is consequently impaired. Also, errors in a quantitative analysis of the components of a mixture may result from preferential vaporisation of the more volatile components of the mixture contained in the needle. This latter problem may be obviated by the use of the *solvent flush technique*. The syringe is repeatedly washed with solvent and then solvent is drawn into the syringe to a reading of 1 μl. The needle is removed from the solvent and the plunger is further pulled back until the total volume of solvent, c. 2 μl, is drawn into the barrel of the syringe and an air space is just visible (Fig. 11a). The desired volume of sample is introduced into the syringe by observing the increase in volume of the air space and the volume is accurately measured by drawing the sample into the barrel of the syringe (Fig. 11b). When the sample is injected in the normal manner, the solvent flushes the sample from the needle and also completely occupies the dead space.

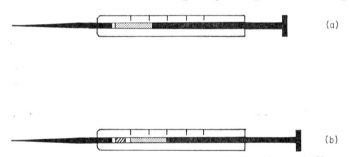

FIG. 11. The solvent flush technique for liquid samples[30].

Several alterations have also been made to the design of the simple syringe which improve the accuracy of injection. One modification involves a precision plunger travelling in the bore of the needle. The plunger comes to the tip of the needle, thus eliminating any dead volume. Best results are obtained when the whole operation of inserting the needle, discharging the sample and the removal of the needle from the injection block is carried out as rapidly as possible. However, even with the utmost care it is generally considered that unless the solvent flush technique is used the highest accuracy attainable in the visible estimation of the amount of sample injected is not better than $\pm 5\%$. As a syringe ages, general wear of the plunger produces errors in the accurate delivery of the sample. The sample escapes past the plunger, being forced back by the high pressure of the carrier gas in the injection block, but a teflon tip/O-ring combination can be fitted at the head of the plunger to minimise this loss. In general the problem of wear of the precision plunger type of syringe is negligible and a greater problem is the clogging of the needle by the sample. Many syringes are demountable but owing to the precision fit of the plunger in the needle, extreme care has to be taken when the syringe is dismantled.

(b) Solid Samples: It is convenient to introduce solids in the form of a solution in a volatile solvent using a syringe as described above. When, however, dilution of the sample would be undesirable, the solid may be introduced directly and syringes similar to those used for the introduction of liquids may be used. The precision plunger, however, differs from that used for liquid samples in that it is longer and protrudes from the needle. It also has a spiral tip (Fig. 12)[1]. The solid dissolved in a volatile solvent is applied to the tip and the solvent is allowed to evaporate. The plunger is then retracted and the needle is inserted into the injection block in the normal way. The tip of the plunger is then held in the heated zone where the solid vaporises.

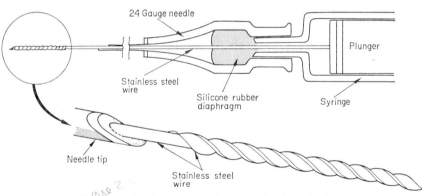

Fig. 12. Modified syringe for the injection of solids.

Several methods are available for the introduction of the solids in which the sample is introduced in an encapsulated form into the injection block. The capsule is usually made of glass[2] which, after introduction into the heated area of the injection block, is broken to liberate the vaporised solid. Indium capsules are also used[3], the sample being released as the indium melts at the high temperature. These methods of introduction have the advantage that the weight of the sample is accurately known. The method by which the capsule is introduced is basically the same for most chromatographs, although the exact design varies according to the manufacturer. The capsule is introduced via an air lock and moved either magnetically or by a probe into the heated zone. To reduce any disturbance of the gas flow, and to minimise the fluctuation in temperature which results from the heat of sublimation of the solid, the sample is allowed to vaporise completely before it is released from the capsule. Using these techniques one may also analyse viscous liquids.

Since with the encapsulation technique both the temperature of the sample and the length of time the sample is held at that temperature can be controlled, it is possible to carry out kinetic studies on the thermal reactions of the sample. Using an extension of this technique it is possible to analyse certain types of non-volatile solids. The sample is pyrloysed at high temperatures of the order of 500 to 800° and the volatile pyrolysis products are then swept onto the column by the carrier gas and analysed in the usual manner. This procedure is known as *pyrolysis chromatography*[4] and has been used extensively in the analysis of high molecular weight organic compounds, in particular for the study of polymers. The chromatogram is often too complex for the identification of the individual peaks but the reproducibility of the pyrolytic decomposition under controlled conditions allows the chromatogram to be used as a "fingerprint" for a given sample. The reproducibility of the "fingerprint" chromatogram, however, appears to be attainable only if a constant pyrolysis temperature is maintained. It has been shown that changes of as little as 25°C in the pyrolysis temperature produce changes in the relative intensities of the peaks of the resultant chromatogram. Larger changes in the pyrolysis temperature can result both in the appearance of new peaks and in the disappearance of other peaks. Many of these differences arise either from incomplete decomposition of the polymeric material or from further secondary reactions of the initially formed products. Other factors, such as the design of the pyrolysis chamber and the catalytic effect of the material from which it is constructed, may also affect the pyrolysis chromatogram. These effects can, to some extent, be counteracted by pyrolysing the sample as close as possible to the pyrolysis chamber outlet so that the initially formed products are immediately swept onto the column.

Figure 13 shows a typical arrangement for a tubular pyrolysis unit which replaces the normal injection port. The pyrolysis chamber is a metal tube which can be rapidly heated to very high temperatures by passing a high current through it at a low voltage, thus ensuring rapid pyrolysis of the sample and vaporisation of the decomposition products. The sample is contained in a boat the temperature of which is recorded by a thermocouple.

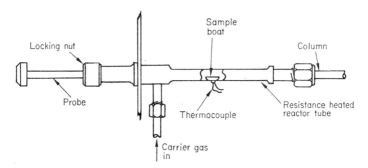

Fig. 13. A flash pyrolysis unit.

Pyrolysis chromatography can be utilised further for elemental analysis. The sample is completely decomposed in the presence of an oxidation catalyst or of oxygen. Under these conditions organic compounds are broken down into carbon dioxide, water and other small molecules such as oxides of nitrogen. The nitrogen oxides are usually reduced to nitrogen over copper prior to their introduction onto the column. In this manner the carbon, hydrogen and nitrogen content of the sample is readily estimated as carbon dioxide, water and nitrogen and commercial analysers are available which can process a complete analysis on a 0·5 mg sample in ten minutes.

(c) *Gas Samples*[5]: Although the use of gas-tight syringes is favoured by many analysts, a preferable and often equally convenient method utilises especially designed gas sampling valves. Gas syringes are capable of withstanding relatively high pressures of the order of seven atmospheres without leakage, but as described in the use of syringes for the injection of liquid samples, the introduction of accurately measured volumes of a gas remains a problem. The inherent errors in this method of introduction lie in the difference in the pressures of the gas in the syringe and in the injection block and also the temperature differences. Both factors affect the calculation of the volume of the gas introduced. Also, a variable and unknown volume of gas contained in the needle may pass into the chromatograph.

In the alternative method, using a gas sampling valve, a known volume of gas at a known pressure and temperature is completely swept into the chromatograph by the carrier gas. Several designs of gas sampling valves have been described. Figure 14 illustrates a simple six way valve. The gas may be introduced at a controlled pressure from a gas line, via the six way tap, into the evacuated sample loop of known volume. The

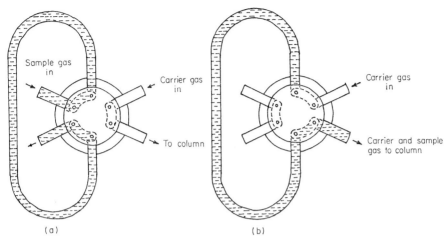

FIG. 14. A six-way valve for the introduction of gaseous samples.

carrier gas is then routed to sweep the sample onto the column without preheating.

Extremely volatile liquids may be introduced in a similar manner, either by condensing the sample in a cooled sample loop (Fig. 15) or in a sample transfer cell of the type shown in Fig. 16, from which it may be introduced onto the column either directly, via the normal injection port, or by purging from the cell with the carrier gas. The condensed sample may alternatively be introduced into the evacuated sample loop of a gas sampling valve of the type shown in Fig. 17.

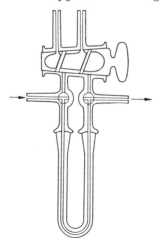

FIG. 15. Liquid nitrogen trap for volatile samples.

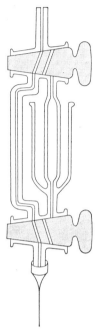

Fig. 16. Sample transfer cell with liquid nitrogen trap.

C. The Oven

In Chapter 1 of this book it was shown that the retention volume of an eluted sample is dependent upon the temperature of the column (see p. 16). It is therefore important that if the retention volume is to be used for characterisation, no fluctuation in temperature should occur along the length of the column during the measurement. To maintain a constant temperature the column is enclosed in a thermostatically controlled oven, the optimum operational temperature of which will be governed by the nature of the sample under investigation (see p. 68) and also by the nature of the stationary phase (see p. 63). For isothermal work an insulated oven constructed of a metal of high thermal capacity can be used. A more efficient method which accurately controls the temperature of the column, but which will also allow the operator to change the temperature conveniently and rapidly for temperature programming (see p. 117), is one in which the column is suspended in an insulated air oven through which the air is circulated at high velocity by means of fans or pumps. Ovens of this type have been designed which can hold the temperature accurately to within $0.1°$ and also allow the temperature to be increased at a rate of 50° per minute and permit equally rapid cooling.

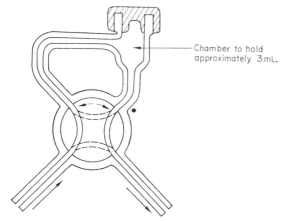

FIG. 17. A by-pass sample valve for use with a syringe.

D. The Column
I. PACKED COLUMNS

Packed columns are usually constructed from glass, copper, stainless steel, or aluminium tubing and those used for routine analysis usually have an internal diameter of between $\frac{1}{16}$ in and $\frac{3}{8}$ in (2 to 9 mm). For preference, glass columns should be used at high temperatures when the metal tubing would catalyse decomposition of the sample and, as noted earlier, if a glass column is used, it is imperative to incorporate a glass liner in the injection port, otherwise decomposition of the sample may occur before it reaches the column.

The liquid stationary phase is absorbed on a porous solid support. The ideal features of this support are that it should consist of inert uniformly spherical particles having a large surface area per unit volume and that it should be mechanically strong over a wide temperature range. The most commonly used supports are diatomaceous earths either kieselguhr, which is sold under the trade names of Celite, Chromosorb W, Embacel and Celatom, or crushed firebrick which has the trade name of Sterchamol, C 22 and Chromosorb P. The crushed firebrick has a higher mechanical strength than the kieselguhr but it has the disadvantage of not being completely inert and will absorb polar samples resulting in tailing of the bands. Undesirable isomerisation and decomposition of the sample is also more prevalent on the firebrick support. Conversely, kieselguhr is comparatively inert but is extremely fragile. With the recent introduction of Chromosorb G, however, the desirable high mechanical strength of the firebrick support and the inertness of the kieselguhr have been combined.

The undesirable characteristics of the firebrick support can be partially attributed to active —Si—OH sites on the surface of the support. Deactivation of these sites may be accomplished by chemical reaction with reactive silyl compounds, such as trimethylchlorosilane (TMCS), dimethyldichlorosilane (DMCS), or hexamethyldisilazane (HMDS), to give inert silyl ethers. For example, the reaction of the active silanol sites with hexamethyldisilazane proceeds in the following manner:

$$\begin{array}{c} —\overset{|}{\text{Si}}\text{OH} \\ \overset{|}{\text{O}} \\ —\overset{|}{\text{Si}}\text{OH} \\ | \end{array} + (CH_3)_3Si.NH.Si(CH_3)_3 \rightarrow \begin{array}{c} —\overset{|}{\text{Si}}.\text{O}.Si(CH_3)_3 \\ \overset{|}{\text{O}} \\ —\overset{|}{\text{Si}}.\text{O}.Si(CH_3)_3 \\ | \end{array} + NH_3$$

The reagent is most conveniently used by injection on to the column at about 80 to 100°C with a slow carrier gas flow rate[6]. The procedure may be repeated until the required performance is obtained from the column. It should not, of course, be used if the solid support is coated with a polar liquid phase the reactivity of which depends upon hydroxyl or other similarly reactive substituents. The alternative procedure involves treatment of the solid support prior to coating with the liquid phase.

Dimethyldichlorosilane reacts with the active sites of the solid support giving both the cyclic silyl ether, formed by the reaction of one molecule of DMCS with two silanol groups, and also the monochlorosilyl ether. Consequently, it is necessary to treat the solid support further with methanol in order to convert the chlorosilyl group into its methyl ether derivative.

$$—\overset{|}{\text{Si}}\text{OH} \xrightarrow{(CH_3)_2SiCl_2} —\overset{|}{\text{Si}}.\text{O}.\overset{\overset{CH_3}{|}}{\underset{\underset{CH_3}{|}}{\text{Si}}}\text{Cl} \xrightarrow{CH_3OH} —\overset{|}{\text{Si}}.\text{O}.\overset{\overset{CH_3}{|}}{\underset{\underset{CH_3}{|}}{\text{Si}}}.\text{O}.CH_3$$

The effect of the particle size upon the column performance has been discussed in an earlier section of this book (see p. 9). A compromise has to be made between the improved column performance which is theoretically possible using smaller diameter particles, and the loss of performance, observed in practice, resulting from the difficulty encountered in packing the smaller particles evenly. It is advantageous, however, whatever particle size is chosen, to use as small a mesh range as possible.

As so many types of column are commercially available, the average organic chemist need not be concerned with packing an individual column. It is often the case that in laboratories which are concerned with the routine analysis of only one or two classes of compounds it may be found that two or three carefully chosen columns will suffice for all separations. In choosing the stationary phase, the two major factors to be considered are (a) the chemical character of the sample to be analysed and (b) the upper and lower temperature limits at which the column may be operated and their relation to the optimum temperature for the separation of the components. These factors, together with the characteristics of the various phases available, are discussed in Chapter 3. A Table of commercially available stationary phases is given in Part II (see p. 153).

However as will be shown in Chapter 3, there are often instances where a custom-made mixed phase column has advantages over a standard single phase column. Also, specialised columns for the separation of, for example, optical isomers are often not commercially available but can be readily prepared in the laboratory. For the preparation of mixed phase columns there is some advantage in using commercially available solid supports already coated with liquid phase. It only requires then to mix the phases in the correct proportions and to introduce the mixture into a column. For simple or specialised single phase columns, the solid support can be coated by one of two procedures. Using the *in situ* method, a solution of the liquid phase is passed through a column filled with the dry solid support. This may be conveniently done using the apparatus shown in Fig. 18. The solvent is removed by blowing nitrogen through the column. This method has the advantage that a coiled column can be used and the procedure therefore may be utilised to recoat old columns. In the alternative method, the solid support is coated prior to its introduction into the column by suspending it in a solution of the liquid phase. The solvent is removed either by evaporation using a rotary evaporator fitted with a fluted flask which by tumbling the slurry during evaporation prevents the formation of lumps, or the slurry may be filtered and then dried in a flow of warm nitrogen.

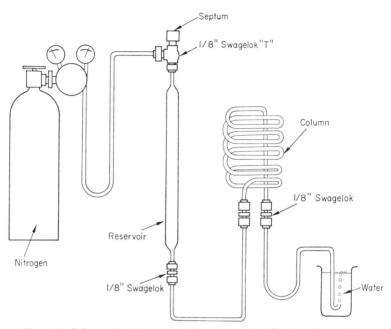

Fig. 18. Schematic diagram of the apparatus for *in situ* coating.

In using precoated solid phases a straight column should be used for the most convenient and efficient packing. This procedure is more suitable for metal than for glass columns, as the metal tube may be coiled in the cold whereas uneven heating of a glass column during coiling will lead to an uneven distribution of the liquid phase. The precoated stationary phase should be sieved to a uniform size and the tubing should be washed with a suitable solvent to remove greases and oils and then dried. One end of the tubing is plugged, preferably with a tight fitting fine mesh screen, and the stationary phase is introduced via a funnel connected to the column by a short length of rubber tubing. Ideally the packing should be uniformly tight with no possibility of channelling. The various vibrational methods used to pack the stationary phase lead to some unevenness (Fig. 19) but it is considered that the best column is obtained by introducing the stationary phase in small amounts and bedding it down by tapping the column on the floor[7]. When the column is filled, a gas pressure is applied to aid the complete settling of the stationary phase. The open end is then sealed with a fine mesh plug. In all of the operations it is important not to crush the fragile solid phase.

A newly prepared column and also those obtained commercially invariably contain volatile materials, either residues of solvent used in its preparation, or adsorbed water vapour. These and other volatile impurities must be removed by heating the column in the chromatograph for several hours whilst a slow stream of carrier gas is passed through it. During the column conditioning process the detector is not connected to the column, but from time to time it is reconnected and the baseline is monitored. When a steady baseline is obtained on the recorder, all the volatile impurities have been removed. It has been advocated, however, that the conditioning process should be continued further until the baseline begins to drift again as a result of volatisation of the liquid phase. The conditioning process has the added attraction of producing more even distribution of the liquid phase on the solid support.

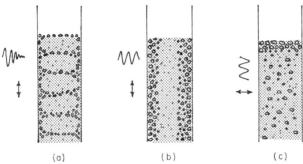

Fig. 19. Segregation of support with different packing methods. a. Addition of packing in small portions bedded down by vertical tapping. b. continuous vertical vibration during continuous addition of packing. c. Horizontal vibration of column during continuous addition of packing.

2. CAPILLARY COLUMNS[8, 9]

Capillary columns differ from the standard packed column in that the liquid stationary phase is distributed as a thin film on the inner surface of a capillary tube. The usual internal diameter of such columns if between 0·25 and 0·5 mm, although both 0·1 mm diameter capillaries for fast analytical work and considerably wider diameter tubes for preparative separation have been described. The main feature of the *capillary column*, or, as it is often called, the *open tubular column* is the low resistance to the carrier gas flow with a consequent low pressure drop along the length of the column. This effectively results in narrower peaks and a higher resolution and also facilitates faster analysis times without loss of column performance (Fig. 20). It also permits the use of considerably longer columns.

Because of the low concentration of the liquid phase in the capillary column, the volume of sample introduced is critical if overloading is not

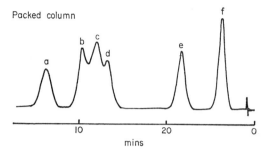
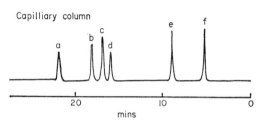

Fig. 20. Comparison of the column efficiencies of a packed and a capillary column. Packed column: 120 × 0·4 cm 20% 7,8-Benzoquinoline on celite at 78·5°. N_2 flow rate 42 ml/min. Column efficiency 1,600 theoretical plates (o-xylene). Capillary column: 12 m × 2·5 mm 12 mg 7,8-Benzoquinoline at 78·5°. N_2 flow rate 0·7 ml/min. Column efficiency 146,000 theoretical plates (o-xylene). a, o-xylene; b, m-xylene; c, p-xylene; d, ethylbenzene; e, toluene; f, benzene.

to occur. This necessitates the use of a stream splitter in the injection block as described on p. 23. Also, as a consequence of the narrow band width and the low concentration of the eluted solute, a highly sensitive detector with a short response time is needed. It was fortunate, therefore, that the capillary column and the flame ionisation detector were developed almost simultaneously.

The performance of a capillary column can be described in terms of the plate height, h, and the average carrier gas velocity, v, by a similar equation to that given for packed columns:

$$h = B/v + (C_g + C_l)v$$

The first term of the van Deemter equation (see p. 8) does not apply to capillary columns. The term B/v describes the band broadening in the gas phase and has the same significance in the above equation for capillary columns as it has in the van Deemter equation for packed columns. The factors C_g and C_l describe the band broadening arising from the finite rate of mass transfer of the solute in the gaseous and liquid phases respectively and can be expressed as $C_g \propto r^2/D_g$ and $C_l \propto d^2/D_l$, where r is the internal radius of the capillary, d the thickness of the liquid film, and D_g and D_l the diffusion coefficients of the solute in the gaseous and liquid phases respectively. Both terms are also functions of the ratio of the liquid to gas phase and the partition coefficient of the solute. Hence the optimum column performance is obtained with narrow diameter capillary and an infinitely thin liquid film. As for packed columns there is an optimum gas velocity and, as the diffusion of the solute in the gas phase is inversely proportional to the viscosity of the carrier gas, one can choose either rapid analysis using a gas of low viscosity, such as hydrogen, or alternatively the use of a gas of higher viscosity, such as nitrogen, at the optimum flow velocity.

The standard capillary column can be modified to give a higher ratio of liquid phase to gaseous phase without further reducing the diameter of the capillary. This is accomplished either by chemical treatment[10, 11] of the internal surface of the capillary, so that it retains a higher proportion of the liquid phase, or by deposition of a support for the liquid phase on the internal surface of the capillary[12, 13]. These porous layer columns combine the high permeability of the simple capillary column with the higher sample capacity of the packed column. It has also been shown that an effective increase in the column performance is obtained as a result of the reduction in the liquid mass transfer term C_l.

Packed capillary columns have also been described[14, 15] which have an internal diameter of <0.5 mm with a solid support particle size of between 200 and 100 μ. The characteristics of these columns are more akin to the simple capillary column than to the packed columns. Their permeability is higher than the standard packed column and they therefore give fast analysis with a high separation power. The sample loading is considerably higher than can be introduced without overloading onto a capillary column and, as a wider choice of stationary phases is available, they have a wide range of application.

E. The Detector

I. INTRODUCTION

The detector senses differences in the composition of the effluent gases from the column and translates the column's separation process into an electrical signal which is communicated via an amplifier to a recorder.

Detectors may be classified into two groups. Those which cause the total quantity of eluted sample to be recorded in an accumulative fashion are called integral recording detectors. The type of record produced by this class of detector is shown in Fig. 21a. The steps in the plot correspond to an eluted compound and the height of each step is a function of the weight of compound. The second class of detector senses the variation in the amount of sample passing through the detector. The record produced takes the form of a differential plot; the signal increasing as the quantity of sample passing through the detector increases to a

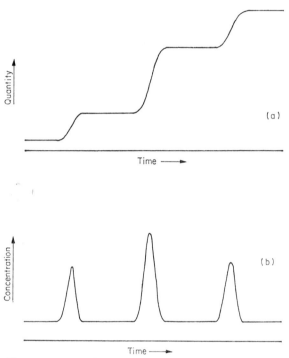

FIG. 21a. Chromatogram given by an integrating detector. b. Chromatogram given by a differentiating detector.

maximum and then falling back to the baseline. This is illustrated in Fig. 21b which shows the elution of the same components with the same relative concentrations as in the integral plot. In the case of the differential plot, it is the area under the peak which is a function of the weight of compound. The response of the differential recording detector may depend either upon the absolute quantity of sample in the detector at any instant or, alternatively, upon the instantaneous concentration of sample in the carrier gas contained in the detector. The former type is sometimes referred to as a *mass detector* whereas the latter type is called a *concentration detector*.

The simple detector system used by James and Martin[16] in their chromatographic separations of monocarboxylic acids and of amines involved passing the effluent carrier gas together with the eluted compounds directly into a *titration cell*. Titration of the eluted acidic or basic compounds was recorded in an integral form. The titration technique has been perfected so as to be fully automatic with a photocell relay system, sensitive to changes in a colorimetric indicator, controlling the addition of the titrant. The system may be used with any of the carrier gases mentioned on p. 54 with the exception of carbon dioxide which would lead to erroneous titration values. On the other hand, detection of permanent gases using a *nitrometer* specifically utilises carbon dioxide as the carrier gas. The volume of the eluted permanent gases are measured by collection over aqueous sodium hydroxide which absorbs the carbon dioxide. The accuracy of the detector is low ($c. \pm 1\%$) particularly when the concentration of eluted sample is small. The titration detector also has a low sensitivity; its optimum sensitivity being attained only when all of the eluted components of the sample lie in a narrow concentration range. The minimum quantity detectable by the titration technique is of the order of 0·01 to 0·02 mg. The system is also restricted in its application and consequently it is rarely used.

The first detector to be designed with a response which was independent of the chemical properties of the eluted sample was based on a simple *gas density balance*[17]. The basic design is illustrated in Fig. 22.

If the effluent gas entering the detector at port D is identical to the reference gas entering the detector at port A, then it will leave the detector by either of the routes DEG or DFG and will not impede the flow of the reference gas. The flow rate past points B and C can be compared by means of thermistors, the resistances of which vary inversely with the logarithm of their temperature. Passage of the gas past the

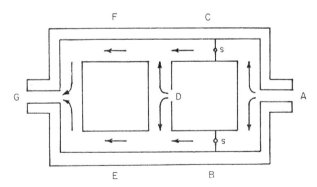

Fig. 22. Gas density detector.

thermistors produces a cooling effect and when the flow rates are equivalent then the cooling effects will be identical. Thus, by incorporation of the thermistors in a standard Wheatstone bridge it is possible to measure any imbalance in the flow rates. If the density of the effluent gas is higher than that of the reference gas, it will preferentially follow route DEG. The passage of the reference gas on the route ABEG is thereby impeded and consequently the flow rate past point C will increase. Conversely, if the density of the effluent gas is less than that of the reference gas the flow rate past point B will increase. Both density differences will produce an imbalance in the Wheatstone bridge but in opposite directions.

The response of the detector is proportional to the density change in the effluent gas and consequently the area, A, under any peak is proportional not only to the concentration of the eluted compound, w, but also to the differences in the densities of the carrier gas and the eluted compound as given in the following equation:

$$A = k.w.\frac{M_c - M_g}{M_c}$$

where M_c and M_g are the molecular weights of the eluted compound and of the carrier gas respectively and k is a constant.

The gas density detector has a sensitivity comparable with that of the thermal conductivity detector (see p. 53). It has the added advantage that the eluted compound does not come into contact with the sensing elements and can therefore be used with corrosive samples. However, although gas density detectors are commercially available they have not found widespread favour.

Numerous other types of detectors have been described. The *hydrogen flame temperature detector*, which has particular application for the detection of combustible organic compounds, utilises the changes in the shape and in the temperature of the hydrogen flame when the gas is contaminated with a combustible compound[18]. The temperature of the flame is measured by a thermocouple situated just above the flame. As an organic compound appears in the carrier gas the flame shape changes and engulfs the thermocouple. The change in the flame temperature is a function of the weight of combustible compound and also of its molar heat of combustion. Thus, although the area of any peak is related to the weight of eluted compound, it is necessary to obtain calibration curves in order to compare the relative concentrations of different compounds.

This type of detector has a high sensitivity and was at one time the popular choice of detector for the routine analysis of organic compounds. It has been superseded, however, by the superior flame ionisation detector (see p. 45).

The adaptation of a simple *ionisation gauge* for use as a sensitive detector for gas chromatographic analysis has been described[19] and is shown in Fig. 23. The electrons from the heated filament are accelerated under the influence of the applied potential and upon impact they ionise the gas molecules in their path. The detector grid is positive with respect to the filament and the applied potential may be adjusted so that the electrons have a lower energy than that required to ionise the carrier gas. The plate has a negative bias with respect to the grid and collects any positive ions formed. The detector therefore responds to any substance having an ionisation potential lower than that of the carrier gas.

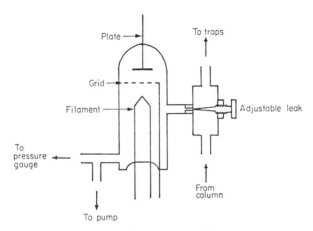

FIG. 23. Ionisation gauge detector.

Because of its high ionisation potential of 24·5 eV, the use of helium as a carrier gas provides one with a universal detector. For use with helium, the applied ionisation potential (i.e. the potential across the filament and grid) is maintained at approximately 18 volts such that no plate current flows in the absence of an organic compound. The sensitivity of the detector is proportional to the gas pressure and over the range 0·02 to 1·5 mm Hg, in which it is possible to obtain a linear response, sensitivities of the order of 10^{-12} mole have been claimed.

Only limited use has been made of *rapid scan spectrometers* (see Chapter 4) to detect gas chromatographic effluent gases. The sensitivity of spectroscopic detectors depends largely upon the molecular structure of the eluted compound. It has been claimed that the limiting sensitivity of ultraviolet spectroscopic detectors for compounds having intense absorption bands can be better than most thermal conductivity detectors and, under certain circumstances, comparable with flame ionisation detectors. The most successful detector of this type is a direct on-line GLC-mass spectrometer combination (see p. 108).

The majority of present day gas-liquid chromatographs are fitted with a flame ionisation detector and/or a thermal conductivity detector. Other detectors, having specific analytical uses, include the electron impact detector and radioactive source ionisation detectors. The design of these detectors depends largely upon the manufacturer and many papers appear in the literature which describe modifications claimed to improve their sensitivity and performance. The following discussion of these detectors describes their basic design and operation.

2. THE THERMAL CONDUCTIVITY DETECTOR OR KATHAROMETER[20,21]

The basic principles of the operation of the thermal conductivity detector depends upon the difference in the thermal conductivity of the carrier gas and a mixture of the carrier gas and the eluted sample. Thus, as most organic compounds have a low thermal conductivity, it is advantageous for high sensitivity to use a carrier gas of high thermal conductivity. For choice, therefore, one would use either hydrogen or helium. However, although it has a thermal conductivity of approximately one sixth that of helium or hydrogen, nitrogen is frequently used because of its cheapness or on grounds of safety. For many purposes it gives satisfactory results, but the lower sensitivity of the detector can lead to concomitant errors of up to 25% in analytical accuracy. A change in the thermal conductivity, resulting from the presence of an eluted sample in the carrier gas, is detected by allowing the gas to flow past a heated filament or thermistor, the temperature of which depends, amongst other factors, upon the ability that the gas has to cool it by conductance of heat. When both the carrier gas velocity and the electrical current applied across the sensor are constant, then the temperature of the sensor is dependent only upon the thermal conductivity of the gas. A change in temperature of the sensor leads to a change in its resistance. Thus, by incorporation of the sensor in a Wheatstone bridge circuit, it is possible to translate a change in thermal conductivity of the gas into an electrical current which can be recorded.

The detector response is proportional to the concentration of the eluted organic compound and its thermal conductivity and also to the filament current but inversely proportional to the flow rate of the gas. The paramount factor upon which the sensitivity of the detector depends is the stability of the temperature of the detector block and, in particular, of the opposing sensors. Highest sensitivity is attained when the difference in the filament and carrier gas temperature is large. Also, as fluctuations in the relative temperatures of the detector block and the carrier gas cause baseline drift and noise on the recorder, it is necessary not only to be able to control the temperatures of the detector and column separately but also to be able to maintain their individual temperatures constant to at least $\pm 0.01°$ and preferably to $\pm 0.001°$. The position of the detector with respect to the column and also the design of the detector block are therefore critical. In practice, the sensors are arranged in pairs; one pair in the reference gas stream and the second pair in the effluent gas stream, as shown in Fig. 24.

In general, the sensitivity of the sensors increase as the filament current increases but, as the base line stability decreases at high filament currents, it becomes a limiting factor to the current which can be used. The extremely large dead volume of the detector, particularly with the filament detector, is also a serious factor limiting the sensitivity which can be attained. This has resulted in the recent introduction of a micro thermal conductivity detector incorporating thermistors.

The connection of the reference and detecting sensors to the gas streams may be made in several ways, the most usual of which are illustrated in Fig. 25.

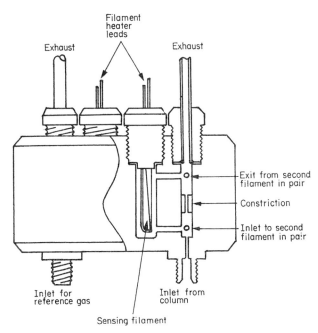

FIG. 24. Hot wire thermal conductivity detector.

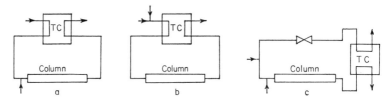

Fig. 25. Schematic diagrams of gas circuits for a thermal conductivity detector.

Arrangement (a) is the most common. The carrier gas first passes the reference sensor prior to the introduction of the sample. The gas re-enters the detector, after having passed through the column, and flows through the detecting circuit. As there is a pressure drop across the column, the flow rates of the gas through the reference and detecting circuits are not equal and, although the recorder can be adjusted to give a steady baseline, this arrangement is susceptible to changes in the flow rate which lead to baseline drift. Arrangement (b) differs from (a) only in the position of the sample injection which now comes before the reference sensor. Other than the production of a peak at the start of an analysis as the unseparated sample passes the reference sensor, this arrangement has little advantage over arrangement (a). The reference sensor in arrangement (c) is supplied with a separate gas flow, the pressure of which is controlled by a valve. This system is consequently less susceptible to flow rate changes.

At high temperatures all the arrangements are susceptible to baseline noise and drift as a result of column "bleeding". This may be obviated by *dual column operation*. The system is similar to arrangement (c) except that the pressure regulator in the reference circuit is replaced by a column identical in its characteristics to those of the analyser column. The use of a dual column system is imperative in temperature programmed analysis.

The thermal conductivity detector is one of the most versatile. It also has the advantage over other detectors in that it is not destructive and the eluted sample can be conveniently collected without the need for a by-pass system before the detector. Compounds which have low thermal stability may of course decompose on the hot wire filament. Tungsten and platinum filaments have a high temperature coefficient of resistance and the filament detector can therefore be used over a wide temperature range. The thermistor detectors, however, are limited in their use to below 100°, as above this temperature their sensitivity falls off rapidly. In using the filament detector it is advantageous, because of its high thermal conductivity, to use hydrogen as the carrier gas, but it is inadvisable to use it with the thermistor detector as there is a high probability that the oxides in the thermistor will be chemically reduced.

3. IONISATION DETECTORS[21,22]

(a) The Flame Ionisation Detector: The ionisation properties of hydrogen: oxygen flames have been known for some time. The observation that the introduction of organic impurities into the flame considerably increased the electrical conductivity led in 1958 to the development of the flame ionisation detector[23, 24]. A typical design is shown in Fig. 26.

Hydrogen is mixed with the effluent gases from the column and burnt in an atmosphere of oxygen. The ions produced in the flame conduct a current from the flame jet, which serves as one electrode, to a second electrode which is situated either directly above or around the flame. The ion current is amplified by a high impedance amplifier and is then transmitted to a recorder. The background current produced by the hydrogen flame may be kept to a minimum by the use of a cool flame obtainable from a low hydrogen to nitrogen ratio. The use of air instead of oxygen also produces a cool flame. The size of the ion current follows Ohm's law when the applied voltage is low and is independent of the distance separating the electrodes. Above *c.* 10 volts/cm, however, the acceleration of the ions due to the applied voltage is at a maximum and the size of the ion current is limited by the ionisation of the particles. The voltage required to maintain this optimum saturation condition is determined by the design of the detector and, in particular, by the distance separating the electrodes.

Since the size of the ion current depends upon the number of ions

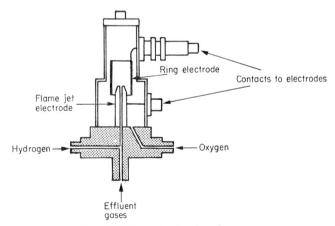

FIG. 26. Flame ionisation detector.

present in the flame, the signal size is proportional to the amount of eluted compound entering the flame in unit time and it is therefore dependent upon the flow rate at which the compound is eluted. The sensitivity of the detector, however, is relatively independent of the absolute flowrate of the carrier gas particularly for those detectors in which the collecting electrode surrounds the flame for, although the height of the flame is approximately proportional to the carrier gas velocity, the diameter of the flame is almost independent of the flowrate.

For optimum performance the shape and size of the hydrogen flame is critical. Thus, for each carrier gas velocity there are optimum hydrogen and oxygen flow rates. The design of the oxygen and hydrogen inlets is also critical. Adequate mixing of the hydrogen with the carrier gas is necessary and is readily attained by means of a simple T-junction. An even concentration of oxygen must be maintained around the flame to prevent "fluttering" of the flame which leads to a high baseline noise on the recorder. Careful purification of the hydrogen and oxygen supplies is also necessary to minimise baseline noise resulting from the presence of organic impurities which may have been introduced into the gases during compression.

The response of the detector to most organic compounds has a linear range of 10^8 and it has been shown to be dependent not only upon the amount of the organic compound in the flame but also upon the number of carbon atoms per molecule of the organic compound[25]. The response differs, however, for hydrocarbons and for compounds containing other elements. Thus, for example, equivalent weights of a hydrocarbon and an alcohol, each having the same number of carbon atoms, will not give signals with the same areas. It is therefore necessary to construct calibration curves in order to deduce relative concentrations directly from peak areas (see p. 80).

The detector is sensitive to c. 10^{-10} g of organic compound, but is either completely insensitive or only weakly sensitive to the inert gases, N_2, H_2, O_2, H_2O, NH_3, H_2S, CS_2, $COCl_2$, COS, CO, CO_2, formic acid, NO, N_2O, NO_2, SO_2, SO_3, volatile silicon compounds and Freons.

(b) The Thermionic (Ion Emission or Electron Impact) Detector: In 1964 Karmen and Giuffrida[26, 27] showed that a relatively simple modification to the flame ionisation detector, involving the introduction of an alkali metal halide probe into the flame, greatly enhanced its response to organo-phosphorus and -halogen compounds. Karmen's detector[26] is shown in Fig. 27. The lower hydrogen flame, into which is introduced the effluent gas from the column, is separated from the upper flame by a grid upon which sodium hydroxide has been deposited. The presence in the

lower flame of organic compounds containing phosphorus, chlorine, bromine, or iodine causes the release of sodium ions from the grid. The upper hydrogen flame acts basically as a thermionic ionisation gauge (see p. 41) which detects the presence of the sodium ions. Thus, if the ion current from this part of the detector is amplified and fed to a recorder, only peaks corresponding to those compounds which contain either phosphorus or halogen will be recorded. It is usual to split the effluent gas from the column and to pass it simultaneously through a

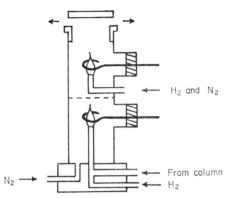

Fig. 27. Two stage hydrogen flame ionisation detector specific for halogens and phosphorus.

thermionic detector and a flame ionisation detector. In this manner a complete record may be obtained of the total composition of a mixture as well as an indication of the phosphorus and halogen compounds.

Considerable advances have been made in recent years in the design of the thermionic detector. Figure 28 illustrates a modern caesium bromide three electrode thermionic detector developed by Pye-Unicam which has a high sensitivity only for phosphorus compounds. A block diagram of the amplifier and recorder circuit is shown in Fig. 29.

The caesium bromide ring acts as the collecting electrode for what is effectively a simple flame ionisation detector and this part of the detector responds to all organic compounds. The presence of phosphorus in the flame produces caesium ions in the flame above the caesium bromide ring. These ions are detected by a third electrode mounted above the caesium bromide ring and concentric with respect to the flame.

In general the range of linear response for the thermionic detector is low, being only 10^3. Its sensitivity, however, is extremely high, the detection limits being between 10^{-10} and 10^{-13} g depending upon the class of compound analysed.

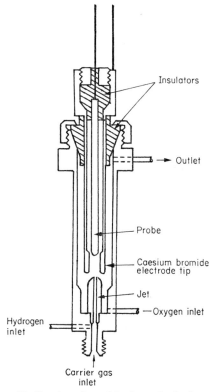

FIG. 28. Caesium bromide thermionic detector.

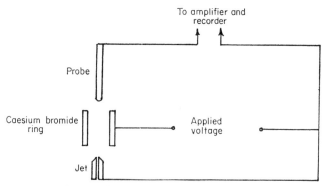

FIG. 29. Circuit for thermionic detector.

c) *Radioactive Source Ionisation Detectors:* As an alternative to the thermal ionisation or flame ionisation detectors, it is possible to ionise the carrier gas by radiation from a suitable radioactive source. In an inert gas the collisions between the electrons produced by ionisation and the inert gas molecules are elastic and they lose little energy. Hence, when a potential is applied across two electrodes placed in the ionised gas, a "standing" current is produced. The introduction of an organic compound into the carrier gas results in inelastic collisions between the electrons and the organic molecules. As a result of these collisions, it is possible for the organic molecules to capture the electrons either to produce ions

$$AB + e^- \rightarrow AB^- + \text{energy}$$

or, if the electrons have sufficiently high energy, the organic molecule may dissociate.

$$AB + e^- \rightarrow A + B^- \pm \text{energy}$$

As a result of these collisions the current is carried by the heavier ions AB^- or B^- which, under the influence of the applied potential, have a lower velocity than the electrons. This leads to a decrease in the size of the "standing" current and, in order to maintain the ion current unchanged, a higher voltage must be applied across the electrodes. Organic molecules which have a strong electron affinity capture the electrons more readily. Hence, the reduction in the ion current is proportional not only to the number of molecules in the detector but also to their electron affinity. Thus, whereas the detector is relatively insensitive to hydrocarbons, it has an extremely high response to halogen compounds (Fig. 30). The electron capture detector may be used, therefore, specifically

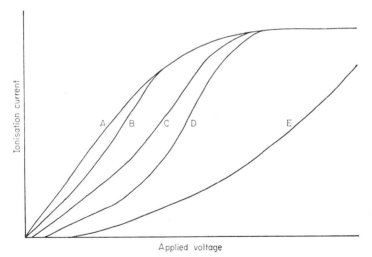

FIG. 30. Plot of ionisation current against applied voltage in an ionisation chamber containing pure nitrogen (A), nitrogen with (B) a hydrocarbon; (C) an ester; (D) an alcohol; (E) a halogenated hydrocarbon.

for the detection of halogen compounds by applying an electrode voltage which is higher than that required to maintain a saturation current for all compounds other than the halogen compounds. The detector may also be used for the analysis of organo-metallic compounds, nitrates, nitriles, and certain sulphur or oxygen containing compounds.

The size of the applied voltage required to maintain the saturation current, as well as depending upon the electron affinity of the compounds in the carrier gas, is also proportional to the distance of separation of the electrodes and depends upon the design and temperature of the detector and upon the pressure of the gas. Since the electron affinities of the molecules vary with the electron energy, it is desirable to maintain the electron energy as constant as possible. This may be achieved by applying the potential across the electrodes in short pulses of c. 1 μsec every 20 to 100 μsec. In this way the electrons interact with the molecules under zero field conditions and only during the 1 μsec pulse are the uncaptured electrons collected at the electrode.

Many radiation sources have been utilised in electron capture and argon detectors. Their suitability depends upon the energy and the range of the emitted particles and, of course, the safety in the operation of the detector is also of considerable importance. γ-Radiation sources are not used for reasons of safety and also because of their non-specific ionisation properties. The radiation from strontium-90 (β-emission) and radium-226 (α-emission), although not presenting a severe health hazard,

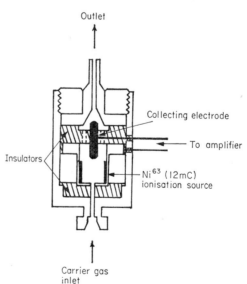

Fig 31. Electron capture detector.

produces an undesirable large background detector current. Of the other suitable sources, tritium in the form of a metal tritide has been extensively used in both electron capture and argon detectors. It produces low energy radiation with a restricted range and therefore does not require special screening. It suffers the disadvantage, however, of releasing tritium gas into the detector effluent at relatively low temperatures. This presents a potential health hazard and special precautions have to be taken in the removal of the exhaust gases. At the present time, the most convenient radioactive source for electron capture detectors is nickel-63 (β-emission). It produces low energy radiation with a short range and has the advantage of a relatively long half life of 85 years. The design of a simple electron capture detector incorporating a ^{63}Ni source is shown in Fig. 31. The usual carrier gas used in conjunction with the electron capture detector is either nitrogen or hydrogen. Argon or helium should not be used alone as their excitation energy is greater than the ionisation energy of most organic compounds. It is possible, however, to use argon as the carrier gas in conjunction with a suitable purge gas, such as methane, when it is then possible to obtain a normal decrease in the standing current as a result of the passage of an eluted compound. This procedure is extremely useful when a dual flame ionisation and electron capture detector system is used.

If argon alone is used as the carrier gas, it is ionised by the β-radiation and collision of an electron with a ground state argon atom produces a metastable excited state (half life of $c.$ 10^{-5} sec.) which can only return to its ground state by transfer of energy via collision with another atom or molecule. As a result of collision between an excited argon atom and an organic molecule, the latter is ionised to yield an electron and a positively charged organic ion. As the excited argon atoms do not contribute to the ion current, the outcome of these collisions is an. increase in the size of the ion current due to the formation of electrons

$$\begin{aligned}&\left.\begin{array}{ll}\mathrm{Ar} + \beta\text{-radiation} \to \mathrm{Ar}^+ + \mathrm{e}^- \ (15\cdot 7 \text{ eV})\\ \mathrm{Ar} + \mathrm{e}^- \qquad\quad \to \mathrm{Ar}^* + \mathrm{e}^- \ (11\cdot 6 \text{ eV})\end{array}\right\} \text{primary process}\\ &\;\mathrm{Ar}^* + \mathrm{M} \qquad\quad \to \mathrm{Ar} + \mathrm{M}^+ + \mathrm{e}^- \qquad \text{secondary process}\end{aligned}$$

Thus the characteristics of a simple electron capture detector change if argon is used as the carrier gas instead of nitrogen. The detector then becomes sensitive to all organic molecules having ionisation potentials of 11·6 eV or less and the ion current increases instead of decreasing as a result of the passage of an eluted sample through the detector. Although once popular, the argon detector is now rarely used. It has a considerably poorer range of linear response than the flame ionisation detector and only a comparable sensitivity.

c

A recent development reported by Brody and Chaney[28] describes the modification of a flame ionisation detector for the semi-specific analysis of organophosphates and of organo-sulphur compounds. These classes of compounds decompose in the flame to give emission spectra with characteristic absorption near 526 nm and 394 nm respectively for the phosphorus and sulphur containing compounds. A *flame photometric detector* (Fig. 32) constructed from a basic flame ionisation detector incorporating a photomultiplier tube with interchangeable filters, transparent only

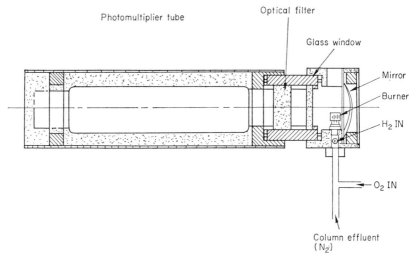

FIG. 32. Flame photometric detector.

to the analytical wavelengths, is commercially available. The detector has a sensitivity to 10^{-9} g and a linear response range comparable to that of the thermionic detector. It has the advantage over the thermionic detector, however, in that it is capable of detecting trace quantities of phosphorus compounds in an excess of organohalides. If the collecting electrode is retained in the flame ionisation detector, simultaneous analysis of both the total composition of the mixture and of the phosphorus and sulphur compounds is possible. The only major drawback to its use appears to be its very high cost.

4. APPLICATIONS

The following Table summarises the limiting sensitivities and the application of detectors in general use. Only a general indication is given of their optimum performance and the reader is advised to consult the manufacturers' specifications for detailed information.

TABLE I
Limiting Sensitivities and Applications of Typical Detectors

	Carrier gas of choice	Minimum detectable quantity (g/sec)	Range of linear response	Application
Concentration Detectors				
Gas density balance	H_2, He (or SF_6)	1×10^{-8}		Universal
Thermal conductivity	He, H_2 (or N_2)	1×10^{-9}	$10^4 - 10^5$	Universal
Electron capture	H_2 or N_2	3×10^{-14}	10^3	Halogeno-sulphur and organometallic compounds
Argon	Ar (or He)	4×10^{-13}	3×10^5	Universal
Mass Detectors				
Hydrogen flame temperature	H_2 or N_2	$c.\ 10^{-7}$	10^4	Universal
Flame ionisation	Ar, H_2 or N_2	3×10^{-12}	10^6	Universal
Thermionic	H_2 or N_2	2×10^{-10}	5×10^2	Phosphorus and halogeno-compounds
Flame photometric	H_2 or N_2	10^{-9}	$c.\ 10^3$	Phosphorus and sulphur compounds

F. The Flowmeter

Several commercial chromatographs are supplied without gas flowmeters and, whilst it is arguable that for routine qualitative analysis of reaction products one need not know the gas flow rate accurately, it is often desirable in comparison measurements to have some measure of the gas velocity. This may be conveniently done by fitting a flowmeter to the exit tube from the column. One of the most simple of flowmeters involves the production of a soap film in the gas stream which acts as a marker. The gas velocity is then determined by timing the passage of the film between calibration marks (Fig. 33). This type of flowmeter is commercially available and is quite inexpensive. Although it is capable of giving very accurate values for the gas velocity, it has the disadvantage of not giving a continuous record. However, for routine analysis this is not too serious a drawback. For continuous recording, one can use a capillary manometer or floatmeter, but as both are susceptible to changes in gas pressure and temperature, they require calibration and ideally they should be maintained under thermostatically controlled conditions.

The effect of the gas velocity upon the column performance is discussed on p. 8.

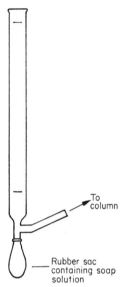

FIG. 33. Soap film flowmeter.

G. The Carrier Gas

The carrier gas should be inert so that it reacts neither with the sample nor with the stationary phase. The selection of a suitable gas for analytical work depends to a large extent upon the operator's requirements in terms of (a) effective separation of the mixture, (b) the analysis time and (c) the choice of detector system. A further consideration in the choice of carrier gas is its availability and its cost, particularly when large volumes are required, as, for example, in preparative chromatography. The commonly used gases are nitrogen, argon, and, to a lesser extent, hydrogen and carbon dioxide. In the United States, where it is relatively inexpensive, helium is widely used. All the gases are available under pressure in cylinders and the hazards involved in their use should not be underestimated. Although the cylinders are pressure tested by the gas suppliers periodically, they should always be handled with care. Whether the cylinders are in use or not, it is preferable that they should be supported in frames by clamps or chains. Correct regulators should always be used. When several instruments are to be operated continuously, thus requiring a large number of cylinders, it is advisable to house the cylinders in a separate reinforced room. In using hydrogen as the carrier gas the exhaust gases should be removed by an efficient ventilation system and it is often advisable, if the chromatograph is part of an industrial product control system, to use a hydrogen generator, several types of which are commercially available, in preference to a cylinder.

For reproducible and reliable results it is imperative that the carrier gas is pure. The major impurities in "pure" gases are water vapour and hydrocarbons. The water vapour can reduce the life of the column and will also increase the baseline noise level on the recorder. The hydrocarbons, which result from the oil pumps used to pressurise the gas, will also contribute to the baseline noise. For routine analysis it is generally sufficient to purify the gas by passing it through a simple molecular sieve drier and a fine filter to remove the oil. Procedures for the preparation of perfectly "clean" gases have been published[29] and commercially produced gas purifiers are available.

References

1. McComas, D. M. and Goldfein, A., *Anal. Chem.*, **35**, 263 (1963).
2. Vandenberg, J. and Kurischka, K., *J. Chromatog.*, **15**, 538 (1964).
3. Nerheim, A. G., *Anal. Chem.*, **36**, 1686 (1964).
4. For a review of this topic see Levy, R. L. *Chromatog. Rev.*, **8**, 49, (1966).
5. Jeffrey, P. G. and Kipping, P. J., "Sample Transfer Systems", Macmillan, New York (1964).
6. Atkinson, E. P. and Tuey, G. A. P., *Nature*, **199**, 482 (1963).
7. Huyton, F. H., van Beersum, W. and Rijnders, G. W. A. *In* "Gas Chromatography, 1960" (R. P. W. Scott, ed.) Butterworths, London (1960).
8. Desty, D. H. *In* "Advances in Chromatography", (J. C. Giddings and R. A. Keller, eds) Vol. 1, p. 199, Edward Arnold, London (1965).
9. Ettre, L. S. "Open Tubular Columns in Gas Chromatography", Plenum Press, New York (1965).
10. Petitjean, D. L. and Leftault, C. J., *J. Gas. Chromatog.*, **1**, 18 (1963).
11. Bruner, F. A. and Cartoni, G. P., *Anal. Chem.*, **36**, 1522 (1964).
12. Halasz, I. and Horvarth, C., *Anal. Chem.*, **35**, 499 (1963).
13. Ettre, L. S., Purcell, J. E. and Belleb, K., *Separation Sci.*, **1**, 777 (1966).
14. Halasz, I. and Heine, E., *Anal Chem.*, **37**, 495 (1965).
15. Halasz, I. and Heine, E. *In* "Advances in Chromatography" (J. C. Giddings and R. A. Keller, eds) Vol. 4, p. 207 Edward Arnold, London (1967).
16. James, A. T. and Martin, A. J. P., *Biochem. J.*, **50**, 679 (1952).
17. Martin, A. J. P. and James, A. T., *Biochem. J.*, **63**, 138 (1956).
18. Scott, R. P. W., *Nature*, **176**, 793 (1955).
19. Ryce, S. A. and Bryce, W. A., *Nature*, **179**, 541 (1957).
20. For a summary of recent advances see Winefordner, J. D. and Glenn, T. H. *In* "Advances in Chromatography" (J. C. Giddings and R. A. Keller, eds) Vol. 5, p. 290, Edward Arnold, London (1968).

21. Svojanovsky, V., Krejci, M., Tesarik, K. and Janek, J., *Chromatog. Rev.*, **8**, 90 (1966).
22. Karmen, A. *In* "Advances in Chromatography" (J. C. Giddings and R. A. Keller eds) Vol. 2, p. 293, Edward Arnold, London (1966).
23. Harley, J., Nel, W. and Pretorius, V., *Nature*, **181**, 177 (1958).
24. McWilliam, I. G. and Dewar, R. A., *Nature*, **181**, 760 (1958).
25. Ongkiehong, L., Ph.D. Thesis, University of Eindhoven (1960).
26. Karmen, A., *Anal. Chem.*, **36**, 1416 (1964).
27. Karmen, A. and Giuffrida, L., *Nature*, **201**, 1204 (1964).
28. Brody, S. S. and Chaney, J. E., *J. Chromatog.*, **4**, 42 (1966).
29. Prescott, B. O. and Wise, H. L., *J. Gas Chromatog.*, **4**, 80 (1966).
30. Kruppa, R. F. and Henly, R. S., *Gas-Chrom Newsletter*, **10**, 1 (1969).

Chapter 3

THE PRINCIPLES OF OPERATION

A. The Analysis of an Unknown Sample 57
   1. The Choice of a Suitable Column 57
   2. Selection of Gas Flow Rate and Column Temperature . . 65
   3. Peak Attenuation 67
B. Pretreatment of the Sample 68
   1. Trimethylsilylation 71
   2. Esterification and Transesterification 73
   3. Acylation 74
References 74

A. The Analysis of an Unknown Sample

For the successful analysis of an unknown sample it is expedient to have a well established routine in order to determine the optimum conditions for separation. The basic sequential procedure for setting up a chromatograph for operation is described in Part II (p. 143). This procedure is, in general, standard for all instruments, although the physical manipulations involved may vary between instruments of different manufacture and are normally well described in the instrument manual. The most important and also the most difficult choice that faces the operator is that of the stationary phase. This, of course, may not be so difficult in the "average" laboratories which possess, at the most, three different types of column. To a lesser extent the correct choice of carrier gas flow rate and column temperature are important in the survey run, as is also the type of detector that should be used in the analysis of a specific sample (see p. 53).

I. THE CHOICE OF A SUITABLE COLUMN

(a) *Initial Considerations:* In choosing the stationary phase one should bear in mind that, the principal objective is to obtain an adequate separation of all the components. It is also advantageous to have, if possible, an adequate distribution of the bands over the chromatogram with each component having an adequate retention volume. Thus, it can be equally worthless to have two components having a hundredfold difference in their relative retention but having an infinitely small retention volume as it is to have two components each with a long retention time but with poor separation. In practice, a compromise has to be reached between the choice of column and the operating conditions and, in particular, considerable use is often made of temperature programmed chromatography (see p. 117). For the optimum separation of the components of any sample, the choice of stationary phase is critical. Primarily one should learn as much as possible about the source of the sample. It is important to know, for example, whether or not the mixture is considered to contain compounds of a similar chemical constitution and also whether the components of the mixture have similar or widely differing boiling points. From a knowledge of the latter information it may be considered advantageous to carry out a prior separation of the higher and lower boiling fractions by simple distillation.

As a general rule, the retention of a polar solute on a polar stationary phase is greater than that of a non-polar solute; the more polar the stationary phase the greater is the difference in their retention. It is also generally true that the relative elution of polar and non-polar solutes from a non-polar stationary phase is a function of the volatility of the solutes, although on the basis of "like adsorbing like" the non-polar solute will be retained by the stationary phase in preference to the polar solute. The order of elution of a homologous series of compounds from a stationary phase, whether it be polar or non-polar in nature, is also a function of their volatility. In using the term, volatility, it is being equated with the vapour pressure of the solute in the gaseous phase at the temperature of the column. It can be shown that, in general, the correlation between the logarithm of the vapour pressure of the solute and its carbon number in any homologous series is linear, and similarly that the logarithms of the relative retention volumes of members of a homologous series are approximately proportional to their boiling points. Each homologous series of compounds gives rise to a separate line (Fig. 34). Deviations from linearity, however, are often observed for many solutes, particularly for the lower members of the series, on both polar and non-polar columns. There appears to be a correlation between this deviation from linearity and the temperature of the column. It has been observed that the deviation is more pronounced for those members of a homologous series whose boiling points are lower than the column temperature. The non-linearity is, in general, more pronounced for non-polar solutes on non-polar columns and similarly for polar solutes on polar columns.

These "rules" are, however, useful only if detailed information of the composition of the sample is known. When no information is available, the stationary phase of choice for the initial chromatogram should be non-polar. The components will be retained approximately in the order of their boiling points and thus the chromatogram will indicate whether fractional distillation is feasible. Without prior knowledge of the source of the sample it would be somewhat premature at this stage to correlate the observed retention data with published data for known compounds.

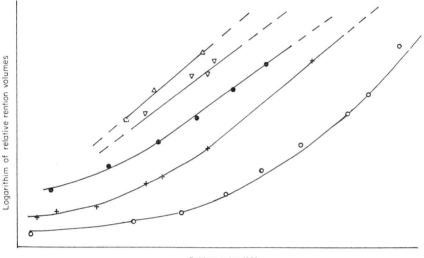

Fig. 34. Plots of the logarithm of relative retentions on 15% LAC2R-446 on Chromosorb W as a function of the boiling point for homologous series. ○ alkanes; + alkenes; ● arenes; △ ketones; ▽ esters.

However, the use of a second column having polar characteristics will present a chromatogram of the components eluted in the order of their increasing polar character (*vide infra*). One can normally identify the components on the two chromatograms by a comparison of peak heights. This procedure must be applied with caution, however, as coincidence of peaks may occur on either column. Also, in the use of the polar column, components having a highly polar character or having groups capable of hydrogen bonding may be retained for an inordinately long time. This will result in band broadening such that it may be difficult to observe the peaks. This is particularly the case when the components are present in low concentration. The retention data from the two chromatograms may provide sufficient information to identify the various types of components present in the sample and will aid the selection of the most suitable column for the analytical separation.

Obviously in the initial survey runs it is desirable to restrict the analysis time to a minimum. This may be accomplished by operating the instrument with the column temperature as high as is feasibly possible (see p. 153).

(b) Polar and Non-Polar Columns: One can classify liquid stationary phases into three groups according to the chemical structure: (a) polar phases which have a high ratio of polar or polarisable groups per molecule, (b) non-polar phases which have no polar or polarisable groups, and (c) semi-polar phases which cover the wide range between these two extremes. However this classification is quite arbitrary. The polarity of the liquid stationary phase may best be defined in terms of its effect upon the retention volume of a polar solute relative to that for a non-polar solute. The greater the polarity of the stationary phase, the greater is the difference in the retention volumes for these two types of solute. In this definition one must appreciate, of course, the distinction between polar and non-polar solutes. Attempts to obtain a more precise definition have entailed comparisons of the retention volumes, obtained using various stationary phases, of solutes of various polarity[1]. In this manner, the polarity of the stationary phase has been expressed in terms of a *polarity index*[2]. This is again an arbitrary scale on which non-polar squalane is given the value zero and β,β'-oxydipropionitrile is the standard polar phase with a relative polarity index of 100. The basis for these measurements is that polar and non-polar solutes of identical boiling points will have identical retention volumes on a non-polar column (assuming that there is no interaction between either solute and the stationary phase), and that the difference in their retention volumes on a polar column will be governed by the degree of mutual interaction between the polar solute and the polar stationary phase (see p. 58). Littlewood[3] has obtained a polarity scale for stationary phases using the basis that the difference in the retention volumes for an homologous series of non-polar solutes, is greater on non-polar than on polar columns (see p. 93). There also appears to be a case for the use of dielectric constants as a direct measure of polarity[4].

Any table of polarity deduced by these methods must be only approximate and will be dependent to a large extent upon the choice of solute used in the measurements. In particular, solutes which are capable of forming hydrogen bonds will behave differently from other polar solutes. A further cause for the variations in the polarity scales results from the mode of interaction between the solute and the stationary phase. The polar character of a molecule is governed by the presence of electron withdrawing or donating substituents and, depending upon the conditions, the molecule may have an electron deficient character, i.e. it is an electron acceptor, or it may have an electron excessive character and is thereby an electron donor. Hence the term "polar", as used previously, is obviously too general and in defining the stationary phase it is preferable to consider electron excessive and electron deficient phases. Similarly the solute may be considered to be electron donating or withdrawing.

In order to obtain a measure of the electron donor–acceptor character of the stationary phase, the ratio of the retention volumes is measured for two "standard" solutes which have, as near as possible, the same boiling point but differ in their electron donor–acceptor character. Such a comparison will distinguish electron donating stationary phases from those which are electron accepting but, to obtain a measure of their relative polarity, it is necessary to introduce data for a non-polar solute. The retention fractions F_n, F_a, and F_d which are obtained from the retention volume data by the following equations

$$F_n = \frac{V_n}{V_n + V_a + V_d}$$

$$F_a = \frac{V_a}{V_n + V_a + V_d}$$

$$F_d = \frac{V_d}{V_n + V_a + V_d}$$

where V_a, V_d and V_n are the retention volumes for the electron accepting, the electron donating and the non-polar solutes respectively, may be used to define the relative polarity as well as the electronic character of the stationary phase[5]. This is demonstrated on the triangular graph (Fig. 35). Brown[5] used n-decane (b.p. 174°) as the non-polar solute and 1,1,2-trichloroethane (b.p. 113°) and dioxane (b.p. 101°) as the electron acceptor and donar solutes. On the graph the points for non-polar stationary phases are located near apex N and polarity increases towards the base DA. Electron donors will lie to the right hand half of the graph with the highly polar electron donating phases located near apex A. Conversely, the most polar electron withdrawing phases are located towards apex D.

From the foregoing discussion it is obvious that in the survey investigation of an unknown sample it would be preferable to measure three chromatograms using a non-polar column and two polar columns, one

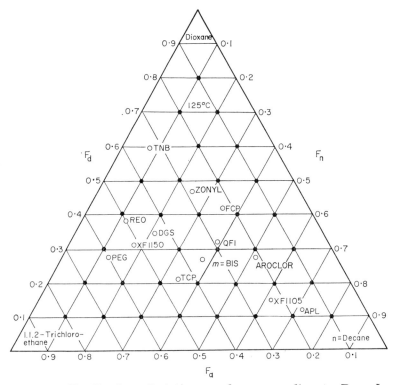

Fig. 35. Classification of stationary phases according to Brown[5]. APL = Apiezon L; XF1105 and XF 1105 = cyanoethylated silicones; AROCLOR = chlorinated biphenyl; m-BIS = m-bis(m-phenoxyphenoxy)benzene; QF1 = fluorinated silicone; TCP = tricesyl phosphate; ZONYL = a perfluoro ester; DGS = poly diethylene glycol succinate; PEG = polyethylene glycol 1500; REO = Reoplex 400; TNB = 1,3,5-trinitrobenzene; FCP = diester of tetrachlorophthalic acid and 2,2,3,3,4,4,5,5-octafluoropentan-1-ol.

being electron accepting and the other an electron donor. Such chromatograms would provide more precise information concerning the structures of the components and further aid the choice of the most suitable stationary phase.

The most commonly used liquid phases are the non-polar silicone gums and oils and the high molecular weight hydrocarbons, the semipolar polyethylene glycols and polypropylene glycols, the polarity of which increases with decreasing molecular weight, and the strongly polar polyesters such as polypropyleneglycol adipate. The glycols are capable also of interaction with the solute via hydrogen bonding. The temperature limits of a selection of these liquid phases and their major applica-

TABLE II

Commonly Used Stationary Phases

Stationary phase	Maximum operational temperature (°C)	Applications
Squalane ($C_{30}H_{62}$)	150	Separation of hydrocarbons
Apiezon (high M.W. hydrocarbon)	< 300	General purpose non-polar phase for separation of high b.p. compounds
Silicone gums (e.g. SE-30)	< 350	General purpose non-polar phase
Di-isodecyl phthalate	150	General purpose semi-polar phase
Polyethylene glycols (e.g. Carbowax 1540)	150	Polar phase suitable for separation of polar compounds, e.g. alcohols, ketones, esters
Polypropylene glycols (e.g. Ucon LB-500-X)	< 200	General purpose phase for the separation of polar compounds
LAC-2R-466 (polydiethyleneglycol adipate cross-linked with pentaerithritol)	200	Polar-phase suitable for the separation of polar compounds

tions are given in Table II. A more detailed description of columns having specific uses is given in Part II (p. 153). In spite of the enormous number of stationary phases available, it is often found that no one phase will effect an efficient separation of a complex mixture. In such instances it may be possible to collect the components which are not well resolved and to rechromatograph them on a different type of column. This procedure is laborious and often, due to the low concentrations of the components, it is also impracticable. Essentially the same procedure may be adopted, however, by the use of a *mixed phase* column. Such a system consists of either two lengths of column containing different stationary phases connected in series or a single column containing a mixture of the two stationary phases[6]. In the latter case, such a column is constructed by coating the solid support *in situ* with the two phases either simultaneously or sequentially, or alternatively by constructing the column from two batches of solid support which have been coated independently with one or other of the liquid phases. The character of the mixed phase column will be governed by the weight per unit length of column of the two stationary phases. Thus, knowing the form of the chromatograms for a sample on the two individual stationary phases, the most suitable combination of the phases may be readily assessed. The usefulness of this technique in the separation of a mixture of methyl iodide, ethyl iodide, cyclohexene, and cyclohexane is illustrated in Fig. 36. The use of Carbowax 400 does not separate methyl iodide and cyclohexene and Silicone Oil does not resolve ethyl iodide and cyclohexane[7]. An acceptable separation is obtained for all four components at a ratio of either 0·14 : 0·86 or 0·86 : 0·14 Carbowax 400 : Silicone Oil.

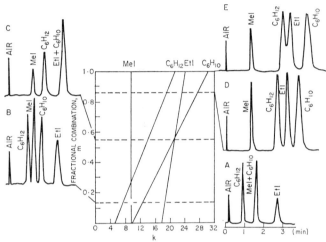

FIG. 36. Dependence of the separation of compounds upon the fractional combination of a two phase column (100 cm). A. 10% Carbowax 400. B. 86·5% of 10% Carbowax 400 + 13·5% of 10% Silicone oil. C. 45% of 10% Carbowax 400 + 55% of 10% Silicone oil. D. 14% of 10% Carbowax 400 + 86% of 10% Silicone oil. E. 10% Silicone oil.

The ratio of the weight of the liquid phase to the weight of the solid phase is not usually critical and proportions of up to 30% have been reported. The most commonly used ratio is 20% by weight. Low ratios of the order of 5% have the advantage of being less susceptible to "bleeding" and also the elution times are usually shorter with a higher column performance than those obtained from the more highly loaded columns. They have the disadvantage, however, that the solute is more readily adsorbed by the solid support resulting in tailing of the eluted bands. Thus, whereas the length of the column does not affect the column performance when there is a high percentage of liquid phase on the solid support (c. 20–30%), it has been found that the column performance is proportional to the column length for the lower ratios of liquid phase (< 5%).

2. SELECTION OF GAS FLOW RATE AND COLUMN TEMPERATURE

After the column has been installed the system should be checked for gas leaks and the carrier gas flow rate then adjusted to the optimum value. A leak check is conveniently accomplished by allowing the carrier gas to flow through the system such that it gives a steady reading on the flow meter which is fitted before the sample port. The gas exit from the instrument is then sealed and, if the system is free of leaks, the flow meter should give a zero flow reading, thus indicating no flow of gas through the system. If the meter fails to give a zero flow reading, one should first ensure that the instrument does not contain a dual or multiple vent, as for example a stream splitting device, before searching for the leak. Usually if the instrument has performed satisfactorily on previous occasions, the most probable sources of the leak are the column fittings. Another common leak source is the septum in the injection port. If the gas leak is not immediately obvious on inspection, it then becomes necessary to locate the area of the leak by isolating sections of the system.

The van Deemter equation (see p. 8) indicates that the efficiency of a column depends in a complex manner upon the flow rate of the carrier gas. The graph derived from the equation (Fig. 37) indicates that there is an optimum flow rate but that at higher flow rates there is only a moderate decrease in the column efficiency. It is therefore advisable to set the flow rate in the survey run as high as is practicable.

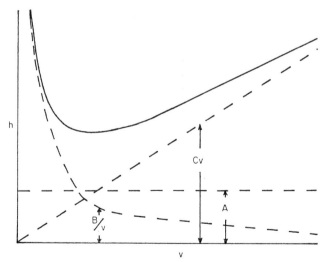

FIG. 37. Graphical representation of the van Deemter equation.

A high flow rate also has the added advantage that the analysis time for the survey runs can be kept to a minimum.

The operating temperatures of columns are defined in specific ranges. As the liquid stationary phase will have a vapour pressure the carrier gas will continuously elute it from the solid support. This is referred to as *column "bleeding"*. Although the vapour pressure of the stationary-phase may be influenced by the choice of solid support and by the load ing on the support, the rate of loss of the liquid phase by "bleeding" correlates with its boiling point and the column temperature. The practical upper temperature limit of the column is usually set therefore at about 150–200°C below the boiling point of the liquid phase. Column "bleeding" is undesirable for obvious reasons. The gradual elution of the liquid phase will change the separation properties of the column until they are no longer acceptable. Eluted solutes will also be contaminated by the liquid phase and thus, as well as disturbing the functioning of the detector, the purity of any trapped solute will be impaired. A secondary factor is the thermal stability of the liquid phase. In several instances polymerisation has been observed at high temperatures and traces of oxygen in the carrier gas have been known to lead to oxidation of the liquid phase at high temperatures.

The lower temperature limit is set by the increased viscosity or solidification of the liquid phase at low temperatures. Both conditions will impair the column performance. There is, however, no simple procedure to determine the optimum column temperature for a complex mixture. The analysis time is minimised by the use of a high temperature but, as with the use of a high carrier gas flow rate, the effective separation of peaks is impaired. Also, at the higher temperatures the thermal stability of the solute can be an important factor. If the column temperature is too low the analysis time can become excessively long and due to diffusion the bands become broadened (see p. 9).

It is often most practical to arbitrarily choose an "average" column temperature and carrier gas flow rate for the initial run and to adjust the parameters in the light of the resulting chromatogram. However, it is never possible to optimise the conditions for all components of a complex mixture and a compromise always has to be made. Although both parameters are readily altered, a change in the column temperature will bring about a more marked effect upon the separation of peaks than will a change in the carrier gas flow rate. As a general rule, a 30° change in column temperature will either double or half the retention volume. It is usual, therefore, to operate the chromatograph at a standard flow rate of, for example, between 25 and 50 ml/min and to adjust only the column temperature. For the optimum efficiency and sensitivity, when any change in column temperature is made, it may also be necessary to alter the operating temperatures of both the sample injection block and the detector (see p. 23). It should also be remembered that it is necessary to allow 10 to 15 minutes for the oven temperature to reach equilibrium although the carrier gas flow rate is usually stable within one minute.

Separation of components may also be improved by using a longer column. Small sample sizes also improve the resolution. The use of a variable temperature programme (see p. 117) may be used with some advantage for the complex mixtures which contain components having a wide range of retention volumes.

3. PEAK ATTENUATION

Further to the information concerning the source of the sample, it is also useful to know the possible number of components in the sample and their relative concentrations. Such information helps the operator not only in ascertaining whether the column has successfully separated all the components but it also aids him in choosing the most suitable settings of the detector and recorder controls and also in his choice of the optimum sample size.

The input potential to the recorder which will produce a full scale deflection of the pen depends upon the type of recorder and it is often usual that under normal operating conditions the passage of a large concentration of component through the detector will produce an output potential from the detector amplifier considerably in excess of the full scale deflection potential of the recorder. Most detectors have a useful linear response over a concentration range of at least 10^5 and it is necessary that the recording system should be capable of matching both the sensitivity and range. This may be accomplished by the incorporation of an attenuator between the detector amplifier and the recorder. The attenuator divides the voltage by fixed steps as shown in Fig. 38. In this manner, for example, both the signal from a low concentration component recorded at an attenuation of $\times 2$ and that from a sample with a tenfold higher concentration can be presented on the same chromatogram (Fig. 39) by changing the attenuation to $\times 64$. When the relative concentrations of the components are not known, the normal procedure is to stabilise the recording system at the lowest attenuation, i.e. highest sensitivity to low output signals. It is then necessary during the running of the chromatogram to watch the recorder constantly and, when the pen approaches full scale deflection, to switch the attenuator to the next higher attenuation. Where the attenuation factors are multiples of 2, such a change in the attenuation will bring the pen to 50% deflection. If the pen continues towards full scale deflection, this procedure is repeated until the maximum is recorded. The attenuation setting must then be restored to the original value.

From the survey run it will be possible to judge the most suitable attenuation settings for the analytical chromatogram.

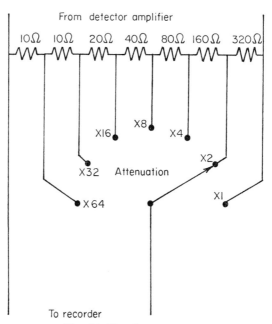

FIG. 38. Simple attenuator.

B. Pretreatment of the Sample

The results of the survey runs may show that, irrespective of the type of column used or of the operational parameters, not all of the components of the mixture are eluted. It may also be found that, as a result of a strong interaction between the stationary phase and the solute, tailing of certain bands is prevalent or that the resultant extended retention time on the column has caused excessive thermal or catalytic decomposition of the solute. These undesirable effects can often be obviated by the conversion of the components into more volatile and usually less polar derivatives. The formation of derivatives can also be useful in the identification of peaks. Thus, for example, the suspected presence of alcohols in a sample may be confirmed by treatment with an acylating agent. Chromatography of the treated sample will show new bands of the corresponding esters having different retention volumes from those of the alcohols whilst the rest of the chromatogram remains essentially the same.

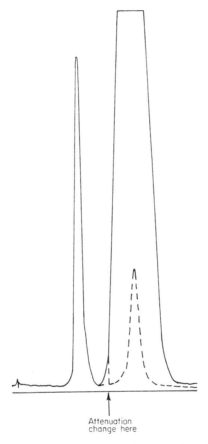

FIG. 39. The effect of attenuation upon the recorded chromatogram of two components of relative concentration 10 : 1 (———— × 2 attenuation, - - - × 64 attenuation).

Most difficulty is experienced in the chromatography of compounds containing carboxyl, hydroxyl, amino, or imino groups. A selection of the most commonly used derivatives for these compounds is given in Table III. For more detailed information of other specific derivatives that have been used the reader is referred to the comprehensive reviews that are available.

TABLE III

Volatile Derivatives of Hydroxy and Amino Compounds for GLC

Class of compound	Reagent*	Derivative	Reference
Hydroxy Compounds			
Carboxylic acids	$CH_3OH : HCl$ or BF_3	Methyl ester	8, 9
	CH_2N_2	Methyl ester	10
	$ClCH_2CH_2OH : BF_3$ or HCl	β-chloroethyl ester (used for low MW acids)	11
Alcohols and Phenols	HMDS + TMCS	Trimethylsilyl ether	12
	BSA or BSFA	Trimethylsilyl ether	13
	Trimethylsilylimidazole	Trimethylsilyl ether	14, 15
	$(CH_3CO)_2O$: pyridine	Acetate	16
	$(CF_3CO)_2O$	Trifluoroacetate	17
	$(CF_3CF_2CO)_2O$	Pentafluoropropionate	18
	Trifluoroacetylimidazole	Trifluoroacetate	19
Carbohydrates	(a) MeOH : HCl (b) HMDS : TMCS	Poly trimethylsilyl ethers of methyl acetal	20
	(a) $(CH_3CO)_2O$: pyridine (b) MeOH : HCl	Polyacetyl esters of the methyl acetal	21
Amino Compounds			
Primary and Secondary Amines	HMDS + pyridine	Monotrimethylsilyl amine	22
	BSA	Mono and bis trimethyl silylamine	14
	CF_3COCH_3	Schiff's base	23
	$(CH_3CO)_2O$: pyridine	Acetyl amino compound	24
	$(CF_3CO)_2O$	Trifluoroacetyl amino compound	23
	$(CF_3CF_2CO)_2O$	Pentafluoropropionylamino compound	23
	Trifluoroacetylimidazole	Trifluoroacetyl-amino compound	19

TABLE III—Continued

Class of compound	Reagent*	Derivative	Reference
Amino acids	(a) (CF$_3$CO)$_2$O		
	(b) CH$_2$N$_2$	Trifluoroacetyl-amino methyl esters	25
	TMCS on Na salt	Trimethylsilylamino trimethylsilyl ester	26
	TMDA	Trimethylsilylamino trimethylsilyl ester	26
Amino alcohols	(a) HMDS+TMCS		
	(b) (CF$_3$CO)$_2$O	Trifluoroacetyl amino trimethyl-silyl ether	12, 23
	(a) HMDS		
	(b) CH$_3$COCH$_3$	Trimethylsilyl ether of the imine	22

* HMDS—hexamethyldisilazane, TMCS—trimethylchlorosilane, BSA—N,O-bis(trimethylsilyl)acetamide, BSFA—N,O-bis(trimethylsilyl)trifluoroacetamide, TMDA—trimethylsilyldiethylamine.

I. TRIMETHYLSILYLATION[27]

The commonly used method for converting alcohols, phenols, glycols and polyhydroxy compounds, steroids, carbohydrates, amines, amino alcohols, and amino acids into their trimethylsilyl derivatives utilises a mixture of trimethylchlorosilane (TMCS) and hexamethyldisilazane (HMDS).

$$RXH + (CH_3)_3SiCl \rightarrow (CH_3)_3SiXR + HCl$$
$$2RXH + (CH_3)_3Si.NH.Si(CH_3)_3 \rightarrow 2(CH_3)_3SiXR + NH_3$$

The reaction is normally completed in 5–10 minutes and, after filtration of the precipitated ammonium chloride, the solution can, in most cases, be injected directly onto the column. Pyridine catalyses the reaction and may be used as a solvent. Its basic properties, however, can produce unwanted side reactions and it will, for example, catalyse the enolisation of steroidal ketones and also the subsequent trimethylsilylation of the enol. Enolisation of the ketone may be minimised by the use of a less basic solvent, or alternatively, the ketone may be converted into its methoxyimino derivative prior to the trimethylsilylation reaction[28].

Other solvents, which have been described for the trimethylsilylation reaction, include dimethyl formamide, dimethylsulphoxide, and tetrahydrofuran. Dimethylsulphoxide, in particular, has been recommended as the solvent for trimethylsilylation of hindered and tertiary alcohols[29]. In a typical reaction c. 10 mg sample in 1 ml dry solvent is shaken at room temperature with 0·2 ml HMDS and 0·1 ml TMCS for 1 minute. The mixture is then allowed to stand for 5–10 minutes to permit the precipitation of the ammonium chloride. Other trimethylsilylating agents which are frequently used include N-trimethylsilyldiethylamine, N-trimethylsilylimidazole, N,O-bis(trimethylsilyl)acetamide and the corresponding trifluoroacetamide derivative. In particular, N-trimethylsilylimidazole appears to be particularly useful for the formation of the derivatives of labile and sterically hindered hydroxyl groups[14, 15].

The reaction of mono- and disaccharides with TMCS and HMDS in pyridine under the conditions described above leads to the fully trimethylsilylated derivatives. These compounds may be used, but it is preferable to prepare the trimethylsilyl ethers of the methyl acetal which is formed by refluxing a solution of the hydroxy compound in 2% methanolic hydrogen chloride for 2 hours[30].

The reaction between amines and HMDS is an equilibrium reaction.

$$2RNH_2 + (CH_3)_3Si.NH.Si(CH_3)_3 \rightleftharpoons 2(CH_3)_3SiNHR + NH_3$$

The reaction is not as rapid as that for the hydroxy compounds so that aminoalcohols initially form the corresponding trimethylsilyl ethers which can be further trimethylsilylated, converted into the Schiff's base with, for example, trifluoroacetone, or acylated. An alternative trimethylsilylating agent for relatively non-volatile amines is N-trimethylsilyldiethylamine. The diethylamine produced in the reaction can be removed by distillation thereby causing the reaction to go to completion.

$$(CH_3)_3Si.N(C_2H_5)_2 + RNH_2 \rightleftharpoons (CH_3)_3Si.NHR + (C_2H_5)_2NH$$

N,O-Bis(trimethylsilyl)acetamide is a considerably stronger silylating agent than HMDS and, as the reaction is non-reversible, it is the reagent of choice for amines. Primary amines form both the mono and disubstituted derivatives extremely readily but the reaction with secondary amines is somewhat slower.

$$2R_2NH + CH_3-C\begin{matrix}O.Si(CH_3)_3\\ \\N.Si(CH_3)_3\end{matrix} \rightarrow 2R_2N.Si(CH_3)_3 + CH_3CONH_2$$

Amino acids have been reported to yield the N-trimethylsilylaminoacid trimethylsilylesters by their reaction with N-trimethylsilyldiethylamine and also by the reaction of the sodium salts of the acids with TMCS[26].

In using the trimethylsilylating reaction the operator should remember that the character of certain stationary phases may be drastically altered as a result of reaction with trimethylsilylating agents. It is therefore not advisable in these circumstances to inject the neat reaction mixture directly onto the column. Also, excessive use of trimethylsilyl derivatives can lead to heavy deposits of silica on the detector, particularly if a flame ionization detector is used. This can lead to a loss of sensitivity.

2. ESTERIFICATION AND TRANSESTERIFICATION

Due to their highly polar character and their ability to enter into hydrogen bonding with the stationary phase, all hydroxy acids must be converted into their corresponding esters. Carboxylic acids are readily converted into their methyl esters by the reaction with methanolic hydrogen chloride or with methanol and BF_3 as a catalyst. The extremely facile and quantitative reaction of carboxylic acids with diazomethane recommends it as the most suitable method for esterification and, as the by-product of the reaction is nitrogen, the reaction mixture may be injected directly onto the column. The reaction, however, has the single disadvantage that, owing to its explosive and toxic properties, the diazomethane must be freshly prepared on each occasion that it is used[31]. Several stable precursors for diazomethane are commercially available. Diazomethane will also convert sulphonic and phosphoric acids quantitatively to their methyl esters and with phenols the methyl ethers are formed. Other methods which have been described include the use of ion exchange resins and the reaction of methyl iodide with the silver salt of the acid[32].

Most low molecular weight carboxylic esters may be chromatographed under normal conditions. However, owing to their involatility, the naturally occurring high molecular weight triglycerides require both a high column temperature and a high carrier gas flow rate and, even when a non-polar column is used, the retention times can be extremely long[33]. Under these conditions thermal decomposition of the sample becomes a serious factor. For satisfactory analysis the glycerides must be degraded into more volatile components. Saponification with alkali followed by esterification of the acids with diazomethane or, alternatively, direct transesterification with methanol and BF_3 or HCl as a catalyst yields glycerol and the methyl esters of the carboxylic acids[34]. The glycerol so formed is not easily chromatographed and in a modified transesterification reaction 2,2-methoxypropane is added to the reaction mixture to convert the glycerol into the less polar isopropylidene derivative[35].

$$\begin{array}{l}CH_2O.CO.R\\|\\CHO.CO.R + CH_3OH + (CH_3)_2C(OCH_3)_2 \xrightarrow{H^+} 3\,RCO_2CH_3 + \begin{array}{l}CH_2OH\\|\\CHO\\|\\CH_2O\end{array}\!\!\!\!\!\Big\rangle C(CH_3)_2\\|\\CH_2O.CO.R\end{array}$$

Mono and diglycerides may be transesterified[34] in the same manner or converted into their trimethylsilyl ethers[36]. Monoglycerides can be converted into volatile allyl esters via their bismethylsulphonyl derivatives[37]. Reduction of mono, di, and triglycerides with lithium aluminium hydride, followed by acylation of the alcohols and glycerol give low molecular weight esters suitable for chromatography[38].

3. ACYLATION

As an alternative to trimethylsilylation hydroxyl, amino and imino groups may be acylated. Acetic anhydride reacts readily with alcohols, phenols and amines in the presence of pyridine to give the corresponding acetyl derivatives. In general, the use of the trifluoroacetyl and pentafluoropropionyl derivatives is to be preferred as their retention volumes are less than the corresponding acetyl and propionyl compounds. This is particularly important in the chromatography of amines, as the retention volumes of the acetyl derivatives are often considerably larger than those of the parent amines. The increased sensitivity of flame ionisation detectors and electron capture detectors to fluoro compounds also recommends the use of the fluoroacyl derivatives in preference to the acetyl compounds. The formation of the fluoroacyl derivatives is extremely facile. The trifluoroacetyl compounds are formed by treatment of the sample with a chloroform solution of trifluoroacetic anhydride at room temperature. The excess anhydride and trifluoroacetic acid is removed by flash evaporation or by extraction with water[17]. For compounds which are labile under acid conditions, the use of trifluoroacetylimidazole and pentafluoropropionylimidazole is recommended.

References

1. see, for example, Ettre, L. S., *J. Gas. Chromatog.*, **1** (2), 36 (1963).
2. Rohrschneider, L., *Z. Anal. Chem.*, **170**, 256 (1959).
3. Littlewood, A. B., *J. Gas Chromatog.*, **1** (11), 16 (1963).
4. Anvaer, B. I., Zhuhkovitskii, A. A., Litovtseva, I. I., Sakharov, V. M. and Turkel'taub, N. M., *J. Anal. Chem. U.S.S.R.*, **19**, 162 (1964).

5. Brown, I., *J. Chromatog.*, **10**, 284 (1963).
6. McFadden, W. H., *Anal. Chem.*, **30**, 479 (1958).
7. Hildebrand, G. P. and Reilly, C. N., *Anal. Chem.*, **36**, 47 (1964).
8. Stoffel, W., Chu, F. and Ahrens, E. H., *Anal. Chem.*, **31**, 307 (1959).
9. Metcalfe, L. D. and Schmitz, A. A., *Anal. Chem.*, **33**, 363 (1961).
10. Vorbeck, M. L., Mattick, L. R., Lee, F. A. and Pederson, C. S., *Anal. Chem.*, **33**, 1512 (1961).
11. Oette, K. and Ahrens, E. H., *Anal. Chem.*, **33**, 1847 (1961).
12. Langer, S. H. and Pantages, P., *Nature*, **191**, 141 (1961).
13. Klebe, J. F., Finkbeiner, H. and White, D. M., *J. Am. Chem. Soc.*, **88**, 3390 (1966).
14. Horning, M. G., Moss, A. M. and Horning, E. C., *Biochem. Biophys. Acta*, **148**, 597 (1967).
15. Horning, M. G., Moss, A. M., Boucher, E. A. and Horning, E. C., *Anal. Letters*, **1**, 311 (1968).
16. Perkins, G. and Folmer, O. F. *In* "Gas Chromatography" (L. Fowler, ed.) Academic Press, New York (1963).
17. Morrissette, R. A. and Link, W. E. *J. Gas. Chromatog.*, **3**, 67 (1965).
18. VandenHeuval, W. J. A., Gardiner, W. L. and Horning, E. C., *J. Chromatog.*, **19**, 263 (1965).
19. Staab, H. A. and Rohr., W., *In* "Newer Methods of Preparative Organic Chemistry", Vol. V., (W. Foerst, ed.) Academic Press, New York p. 61 (1968).
20. Hedgley, E. J. and Overend, W. G., *Chem. Ind. (London)*, p. 378 (1960).
21. Jones, H. G. and Perry, M. B., *Can. J. Chem.*, **38**, 388 (1960).
22. Capella, P. and Horning, E. C., *Anal. Chem.*, **38**, 316 (1964).
23. VandenHeuval, W. J. A., Gardiner, W. L. and Horning, E. C., *Anal. Chem.*, **36**, 1550 (1964).
24. Brooks, C. J. W. and Horning, E. C., *Anal. Chem.*, **36**, 795 (1966).
25. Cruickshank, P. A. and Sheehan, J. C., *Anal. Chem.*, **36**, 1191 (1964).
26. Von Rühlmann, K., and Giesecke, W., *Angew. Chem.*, **73**, 113 (1961).
27. For a general account of trimethylsilylation see Birkofer, L. and Ritter, A. *In* "Newer Methods of Preparative Organic Chemistry", Vol. V, (W. Foerst, ed.), Academic Press, New York, p. 211 (1968) and Pierce, A. E., "Silylation of Organic Compounds", Pierce Chemical Company, Rockford, Illinois (1969).
28. Gardiner, W. L. and Horning, E. C., *Biochem. Biophys. Acta*, **115**, 524 (1966).
29. Friedmann, S. and Kaufman, M. L., *Anal. Chem.*, **38**, 144 (1966).
30. Neely, W. B., Nott, J. and Roberts, C. B., *Anal. Chem.*, **34**, 1423 (1962).
31. Ardnt, F. *In* "Organic Syntheses" Coll. Vol. II (A. H. Blatt, ed.) p. 165 (1943); Redemann, C. E., Rice, F. O., Roberts, R. and Ward, H. P. In "Organic Syntheses" Coll. Vol. III (E. C. Horning, ed.) p. 244 (1955); de Boer, T. J. and Backer, H. J. *In* "Organic Syntheses" Coll. Vol. IV (N. Rabjohn, ed.) p. 250 (1963). Wiley, New York.
32. Gehrke, C. W. and Goerlitz, D. F. *Anal. Chem.*, **35**, 76 (1963).
33. Patton, S., *J. Dairy Sci.*, **43**, 1350 (1960).
34. Kaufmann, H. P. and Mankel, G., *Fette, Seifen, Anstrichmittel*, **65**, 179 (1963).
35. Mason, M. E. and Waller, G. R., *Anal. Chem.*, **36**, 583 (1964).
36. Wood, R. and Snyder, F., *Lipids*, **1**, 62 (1966).
37. McInnes, A. G., Tattrie, N. H. and Kates, M., *J. Am. Oil Chemists' Soc.*, **37**, 7 (1960).
38. Horrocks, L. A. and Cornwell, D. G., *J. Lipid Res.* **3**, 165 (1962).

Chapter 4

INTERPRETATION OF THE CHROMATOGRAM

A.	Quantitative Analysis	77
	1. Introduction	77
	2. Measurement of Peak Areas	81
	3. Measurement of Peak Heights	87
	4. Non-Gaussian Peaks	88
B.	Qualitative Analysis	91
	1. Use of Retention Data	91
	2. Functional Group Analysis	95
	3. Thin Layer Chromatography	96
	4. Physical Methods of Analysis	99
References		114

A. Quantitative Analysis

1. INTRODUCTION

One of the most important features of gas chromatography is its ability to give a quantitative measure of the relative concentrations of components in a mixture. It is also possible, by the use of standards, to obtain reasonably easily a measure of the absolute concentrations.

Detectors may be classified into two types: (a) mass detectors, which give a signal proportional to the mass of the solute reaching the detector in unit time (b) concentration detectors, for which at any instant the signal is proportional to the concentration of the solute per unit volume of carrier gas. The two systems differ in that the signal recorded at any instant from a concentration detector is independent of the mass flow rate whereas that from a mass detector is proportional to the velocity of the solute but independent of the concentration of the solute in the carrier gas. The effect of a change in the carrier gas flow rate upon the signal from the two classes of detector is shown in Fig. 40. As a result of the increased flow rate, the rate of elution of the component will increase and produce a decrease in the recorded peak width. However, whereas the signal from a mass detector increases proportionately with the carrier gas flow rate thereby maintaining a constant area under the peak, the

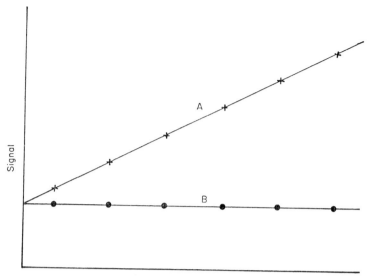

Fig. 40. Dependence of detector signal for a constant sample concentration in the carrier gas upon the carrier gas flow rate for (A) a mass detector and (B) a concentration detector.

signal from a concentration detector remains constant with a consequent decrease in peak area. Thus, for a concentration detector, only if there are no fluctuations in the carrier gas flow rate during the measurement will it be possible to compare peak areas for different components. For mass detectors the area under the peak, A, is proportional to the mass, m, of the component.

$$A = fm$$

The proportionality factor, f, is called the *detector response factor* and its value will depend not only upon the type of detector but also upon the chemical character of the component.

To a first approximation, the composition of a mixture can be readily established by direct comparison of peak areas. The relative percentage concentration of the nth component may be correlated roughly with its relative peak area, A_n, compared to the total area of all peaks.

$$A_1 + A_2 + \ldots A_n = A$$

$$A_n\% = A_n \cdot 100/A$$

For closely related compounds one can assume that the detector response factor would be almost identical for each component and hence the percentage composition of the mixture by weight correlates directly with the relative percentage peak areas. This is not always true for the lower members of a homologous series and the accuracy of the assumption can be checked by comparison of the chromatogram of the sample mixture with that of a standard mixture prepared using the calculated relative concentrations. This simple analytical procedure is usually a sufficiently good indication of the composition of a sample for most general purposes. As an approximate rule for flame ionisation detectors the peak areas of hydrocarbon components may be corrected by multiplication by a correction factor, K, which depends upon the percentage by weight of carbon in the hydrocarbon[1]. The rule has also been applied, with modification, to several other classes of organic compounds[2].

$$K = \frac{\text{molecular weight of hydrocarbon}}{\text{number of carbon atoms per molecule} \times 12}$$

For a more accurate evaluation of the percentage composition of a mixture, the detector response factor for one component may be arbitrarily assigned the value 1·00 and *relative response factors* based on this value may be calculated for the other components. This determination of the relative response factors, which is known as *internal normalisation*, requires accurate values of the peak area for known amounts of pure samples of each component. The areas of peaks produced by equal amounts of two components 1 and 2 are related by the expression

$$A_1/f_1 = A_2/f_2$$

where A_1 and A_2 are the areas and f_1 and f_2 are the response factors of the two components. If component 1 is the reference component, then the relative response factor for the second component is given by

$$f_2 = A_2/A_1$$

The area of each component is multiplied by its relative response factor to give the corrected areas and by a similar treatment to that described for the uncorrected areas, the true percentage composition of the mixture may be calculated.

Standard amounts of each component can be introduced into the chromatograph by injecting, as accurately as possible, known volumes of standard solutions of the components. In using this method it is important that the chromatographic peak of the solvent does not overlap that of any of the components. This technique can be highly accurate when a syringe which has the plunger in the needle is used, but in order to attain any degree of accuracy with a simple syringe the solvent flush technique described on p. 25 must be used. An alternative, but somewhat more tedious, method for the measurement of the amount of sample introduced onto the column involves weighing the syringe before and after injection. As this method also requires measurement of the density of the solution, its overall accuracy is probably no greater than that of direct visual measurement of volume.

The above procedures are only of value if all the components of the mixture are eluted. A common alternative technique is the calibration of the areas of component peaks against that of an added internal standard[3]. In a similar manner to that given above, one can deduce that

$$m = \frac{A}{A_{\text{std}}} \cdot \frac{m_{\text{std}} f_{\text{std}}}{f}$$

where m is the amount of the component under investigation which gives a peak of area A and m_{std} is the amount of standard compound which produces a peak area A_{std}; f_{std} and f are the detector response factors of the standard compound and component respectively.

Reference solutions containing constant concentrations of the internal standard and varying concentrations of the component are prepared and a calibration curve (Fig. 41) is constructed for the ratio of the areas of the component and internal standard (A/A_{std}) against the concentration of the component (m). With this method it is not necessary to calculate the relative response factors, as it is possible by direct comparison of the relative areas of the component and internal standard in the chromatogram of the mixture to determine an accurate value of the absolute

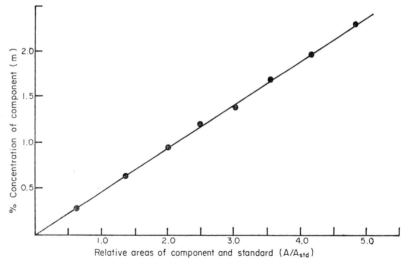

FIG. 41. Calibration curve for the determination of the concentration of a component using an internal standard.

concentration of the component. The accuracy of this method is of the order of $\pm 5\%$. The choice of the internal standard to be used depends entirely upon the composition of the mixture to be analysed. Obviously it must not be a substance which is already present in the mixture and its retention volume must be such that it does not overlap any component peaks. In determining the concentration of any one particular component of a mixture, it is advisable that the retention volume of the internal standard should be in close proximity to that of the component being analysed and that for high accuracy the ratio of the areas of the two peaks should be close to unity. For maximum accuracy two or more internal standards should be used.

2. MEASUREMENT OF PEAK AREAS

Several widely different techniques for the measurement of peak areas are available. A recent study[4] has shown that, although the degree of accuracy of the different methods may differ, there is little to choose between the precision of the measurements. It is therefore a matter of personal choice as to which method is adopted. Of the six methods described in the following sections those using either an automatic integrator or a planimeter and the height × peak width at half height method are the most generally adopted.

(a) Weight of Cutout Peak: This method relies upon the chart paper being of uniform weight and its accuracy depends upon the precision with which the peaks can be cut out. This method was not examined by Mefford *et al.*[4]. Although this technique may be used with some advantage for poorly shaped broad peaks, it is obviously inadequate for sharp narrow peaks.

(b) Height × Peak Width at Half Height: The use of this method in which the peak height is multiplied by the peak width at half height assumes that the peak has a Gaussian shape (Fig. 42) the area of which is given by the expression
$$A = 2 \cdot 507 h \sigma$$
where h is the peak height and σ the standard deviation. As the width of the peak at half height, $W_{h/2}$, is equal to $2 \cdot 354\ \sigma$, it is seen that this method gives a good approximation (94% accurate) to the true area.
$$A = 1 \cdot 06 h W_{h/2}$$
i.e. $\quad 0 \cdot 94 A = h W_{h/2}$

The method requires extrapolation of the base line and the measurement of two readily attainable values. The areas of peaks which have required

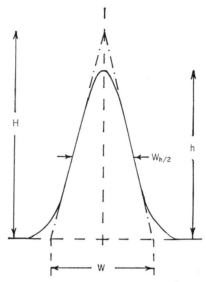

FIG. 42. Plot of Gaussian peak.

attenuation (see p. 67) are also easily determined. The method is limited only in the case of very sharp and intense peaks and for those peaks which have a long retention volume and are consequently usually broad and are of low peak height. The accuracy by which one can measure the half height peak width of such peaks is low. Under certain conditions it is also possible to estimate the area of non-Gaussian peaks by this method[5] (see p. 88).

An alternative form of this method provides a more accurate value of the areas of very narrow and very broad peaks. The width of a Gaussian peak measured at different heights can be expressed in terms of the standard deviation, as shown in Fig. 43. The standard deviation can therefore be determined by direct measurement and incorporation of the value in the above equation gives the peak area. For sharp peaks the standard deviation is most readily obtained from the peak width at the peak height $0.135\ h$, whereas for broad peaks with a low peak height it is more accurate to measure the band width at $0.882\ h$ (see Fig. 43).

(c) *Height × Retention Time:* The increase in peak width resulting from an increase in retention of the solute by the stationary phase has been shown in practice to be linear for many systems, i.e. $t_R = K\sigma$. Consequently it is possible to calculate relative areas by multiplying the height of the peaks by their retention time.

$$A = \frac{2.507}{K} h t_R$$

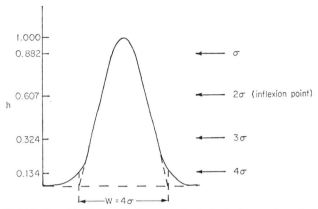

FIG. 43. Relationship between the standard deviation and peak height for a Gaussian peak.

This method is as suitable for the narrow peaks which have short retention times as it is for the broad peaks with long retention times. However, as the relationship between the standard deviation and the retention time has no theoretical basis and as it is often found that it is not always linear, it is advisable to construct a calibration curve of σ (measured in minutes) against the retention time (minutes) for each system investigated[6].

(d) *Triangulation*: For Gaussian shaped peaks the area of the triangle enclosed by the base line and the tangents at the points of inflexion, $H.W_{h/2}$ (Fig. 42), is proportional to the true area of the peak and may be used in the determination of the relative percentage composition of a sample. The method has the same limitations as for method (b) and also the added inaccuracy involved in the construction of the triangle.

(e) *Planimetry*: A planimeter is a commercially available mechanical instrument used to measure the area of any irregular shape. The base line of the chromatogram is extrapolated under the peak and the pointer attached to a movable arm on the planimeter, is carefully traced around the enclosed area. As the pointer traverses the chart it causes a dial and vernier drum to rotate. The difference between the initial and final readings on these scales gives the area of the peak. Fuller instructions for the use of the planimeter will be found in the manual accompanying the instrument. For maximum accuracy the area measurement should be repeated several times and the values averaged. The use of a planimeter is time consuming but the method has the advantage that it can be used with equal accuracy for the measurement of the areas of both Gaussian and skewed peaks. In general, the overall precision and accuracy of the method is high but, as the accuracy of measurement is dependent upon the area measured, the error can be as high as $\pm 5\%$ for small peaks.

(f) *Automatic Integrators:* The current trend is towards fully automatic peak integration. Several different types of automatic integrators are commercially available and, in general, their cost is proportional to their versatility. The least expensive is the mechanical "ball and disc" integrator which operates from the output of the recorder servo system, i.e. the input signal, and is also coupled to the chart movement, i.e. the

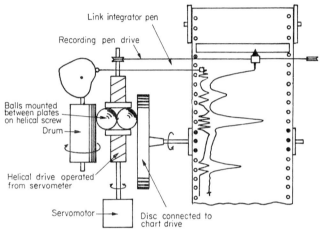

FIG. 44. A ball and disc integrator.

peak width (Fig. 44). The area of the peak is proportional to the product of these two parameters and is obtained by counting the total number of completed and partially completed oscillations which are recorded directly on the chart paper (Fig. 45). As the integrator records only the area of the peak drawn on the chart paper, it is necessary to adjust the attenuation to that most suitable for the recording of each peak and then to correct the recorded area by multiplication with the attenuation factor. Logarithmic recorders and integrators are also commercially available but often their range is limited.

In general, the "ball and disc" integrator is capable of high precision and accuracy but it requires careful adjustment and periodic checks to realise maximum performance[7]. It is difficult, however, with this form of integrator to assess with accuracy the area of overlapping peaks and the measurement of the areas of narrow closely spaced peaks usually

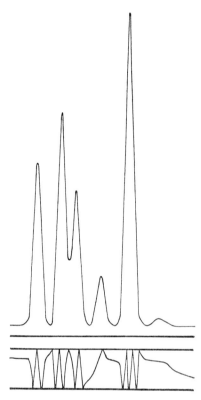

FIG. 45. Chromatogram with recorded integration.

requires that the chart speed be increased in order to obtain high accuracy.

The alternative form of automatic integrator is the electronic digital integrator[8]. Such an instrument can be coupled with a computer and in most instances its cost is considerably in excess of the cost of the chromatograph. Depending upon the degree of sophistication, the electronic integrator is capable of giving in a printed form the retention time and the area of each peak and the summation of the areas. When the integrator is coupled with a computer it is also possible to obtain the percentage composition of the total sample, the individual response factors, the percentage error of each measurement and the most probable identity of each peak of the chromatogram.

The input of the digital integrator is connected directly to the output of the detector, thus making it independent of the recorder. A voltage to frequency converter converts the signal from the detector into a series of pulses whose instantaneous frequency is proportional to the amplitude of the signal and these pulses are counted on an electronic counter. The integrator detects the peak by measuring the rate of change of the output signal from the detector, i.e. it measures the slope of the curve (Fig. 46a). At the start of the peak the first differential of the detector output signal becomes positive and at this point the electronic counter operates and continues to do so until the differential signal changes from a negative value to zero. At this point the total number of counts corresponds to the area of the peak and is printed out. The retention time is

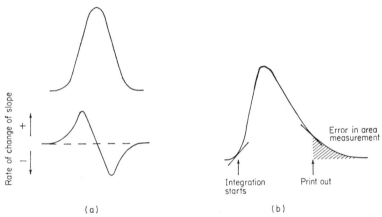

FIG. 46a. Plot of first differential of detector signal against time corresponding to the Gaussian peak. b. Plot indicating error in area measurement of an asymmetric peak when integration starts and finishes at equivalent upward and downward gradients.

recorded when the first differential of the signal changes from a positive to a negative value. The accuracy of this method of peak detection, and ultimately in the accurate measurement of the peak area, depends primarily upon the sensitivity of the instrument to the rate of change of the signal. The accuracy also depends upon the shape of the peak and upon the stability of the base line. With asymmetric peaks the slope of the curve at the start of the leading edge will differ considerably from that at the end of the trailing edge (Fig. 46b). If, therefore, the control gate on the electronic counter operates on equal and opposite values of the slope, an inaccurate area will be recorded. This problem can be overcome by separate slope sensitivity controls for the "up" and "down" gradients. The problem of inaccuracies resulting from a drifting base line can be solved electronically by a system which constantly monitors the base line preceding and following the peaks.

The outstanding advantage of a digital integrator over other types of integrator is its capability to separate overlapping peaks into two separate areas. It automatically detects the point of inflexion or valley between the unresolved peaks and at this point prints out the area of the first peak. The areas obtained in this manner are not grossly inaccurate as shown by the figures in Table IV for the measurement of two overlapping Gaussian curves (Fig. 47).

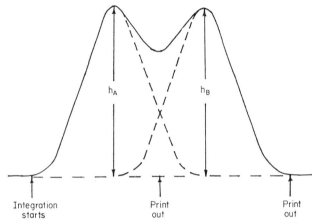

FIG. 47. Integration of two overlapping Gaussian peaks.

TABLE IV

*Errors in Peak Areas for Overlapping Gaussian Peaks**

Ratio of peak heights (h_A/h_B)	Error in area (%)	
	Peak A	Peak B
1	0·2	0·2
2	0·2	0·5
3	0·2	1·5
5	0·2	5·0
10	0·2	10·0

* Data taken with permission from "The Practice of Gas Chromatography" edited by L. S. Ettre and A. Zlatkis. John Wiley, New York, (1967).

3. MEASUREMENT OF PEAK HEIGHTS

An alternative and somewhat less tedious method for the estimation by manual means of the composition of a sample is a direct comparison of the peak heights. The method, however, is not satisfactory when one requires high accuracy. As shown in Fig. 40, the output signal from a mass detector is influenced by the carrier gas flow rate and consequently, unless the operational conditions can be stabilised throughout the chromatographic measurement, the accuracy of the measurements is impaired. The sample size can also affect the mass/peak height ratio. At low concentrations the peak height may give a low value due to loss of sample by adsorption in the chromatographic system and at high concentrations the possibility of overloading of the stationary phase will produce band broadening and consequently an inaccurately low peak height. This problem can be surmounted by the construction of calibration curves for each component. However, although the individual calibration curves will improve the accuracy, it is preferable to calibrate with standard mixtures of the components. A mutual interaction between two components can produce an increased solubility of one of the components (see p. 12) which will lead to an increase in its retention volume and a decrease in its peak height due to band broadening. The accuracy of the technique is further improved by the use of an internal standard in a similar manner to that described for the measurement of peak areas. Any change in the operational parameters will affect both the internal standard and the components to an equal extent. Using these calibration procedures the overall accuracy of the direct measurement of peak heights can be comparable with that of area measurement by the peak height × width at half peak height method. The major error in the area determination results from the measurement of the width at half peak height and for the extremely narrow bands, which are produced by low boiling components at the beginning of a chromatogram, the peak height measurement can have a superior degree of accuracy over the area measurement.

4. NON-GAUSSIAN PEAKS

The areas of non-Gaussian peaks are best determined with a planimeter or with an automatic integrator. If, however, neither of these methods are available, it is still often possible to obtain a fairly accurate value of the area using modifications of the height × width at half height method described earlier.

(a) *Unsymmetrical Peaks:* The area of a skewed peak (Fig. 48) is obtained by the product of the peak height, AB, and the width, EF, at half height, i.e. when AC = CD. This method is only applicable with any degree of accuracy when the locus of the midpoints, ACD, of the peak is a straight line[7].

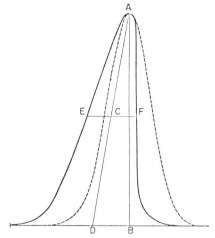

Fig. 48. Area measurement of a Non-Gaussian peak.

(b) *Peaks on Sloping Base Lines:* The area of a peak which has a sloping base line can be measured in much the same way as that described for the unsymmetrical peak. A tangent to the peak is drawn parallel to the sloping base line (Fig. 49) and the line AB constructed along the locus of the midpoints. The width at half height, AC = CB, is obtained by drawing the line DCE parallel to the base line. The area is approximately equal to AB × DE[9].

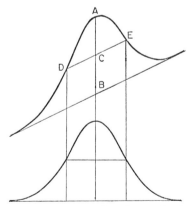

Fig. 49. Area measurement of peaks on sloping baselines.

(c) *Unresolved Peaks:* Only an automatic integrator can determine the area of overlapping peaks with any accuracy. The manual methods will, however, give approximate values of the areas.

Initially one must separate the composite area into two areas. This can be done in three ways. In the first method, which is the same as that used by an automatic integrator, the peaks are divided along the line bisecting the valley and perpendicular to the base line (Fig. 50a). The two areas can only be determined manually by the use of a planimeter and the accuracy is poor for unresolved shoulders. In an alternative method the more intense peak is resolved by construction of the line AB

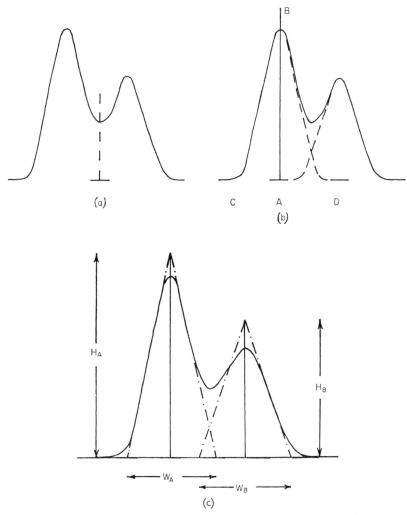

Fig. 50. Area measurement of unresolved peaks a. without resolution, b. by constructional resolution, c. by triangulation.

perpendicular to the base line and passing through the maximum. The total peak, CBD, is then drawn by reflection of the curve CB about this line (Fig. 50b). The second peak is then obtained by subtraction of that part of the peak CBD which lies in the remaining area. The areas of both peaks may be determined by any one of the manual methods described in the earlier sections. By this method both partially resolved peaks and peaks with unresolved shoulders may be separated, but the accuracy of the method depends upon the initial location of line AB. A more simple method which can be applied for partially resolved peaks involves the construction of tangents to the leading and trailing edges of the composite peak and the construction of triangles as shown in Fig. 50c. The accuracy of this technique is poor and should only be used when no other method is suitable.

The areas of small peaks on the trailing edge of larger peaks are difficult to measure and are best determined by treating them as peaks on sloping base lines (*vide supra*).

B. Qualitative Analysis
I. USE OF RETENTION DATA

It has been indicated in Chapter 1 that the retention time is characteristic for a compound and depends upon its solubility in the stationary phase. It would therefore be reasonable to infer that one could identify the components of a mixture by direct comparison of their retention data with that obtained for standard compounds or from published data. However, as has been shown in earlier sections, the retention time is subject to variation and its value is dependent upon column temperature and the carrier gas flow rate, both of which may alter between the calibration and analytical measurements. As the retention time is also dependent upon column length and the type and loading of the stationary phase, any direct comparison with literature data could be erroneous. A more reliable correlation can be obtained by the use of relative retention data, preferably relative retention volumes (see p. 17).

The fact that a correlation exists between the observed relative retention volume and that obtained for a known compound does not, however, constitute proof of identity. The observed peak could be that of another compound which by coincidence has the same retention volume or it could be a composite peak of two or more components having very similar retention volumes and together giving the appearance of a single peak. Thus, the practice of adding a drop of the pure compound to the mixture and observing an increase in the area of the peak at the expected retention volume is also inconclusive. If, however, the added compound gives an extra peak with a retention time different from those of the components of the mixture then one has conclusive evidence for the *absence* of that compound in the mixture. A comparison of the peak width at half height with that of the pure standard compound will confirm whether the peak is a composite one and by lowering the column temperature it may be possible to separate the individual components.

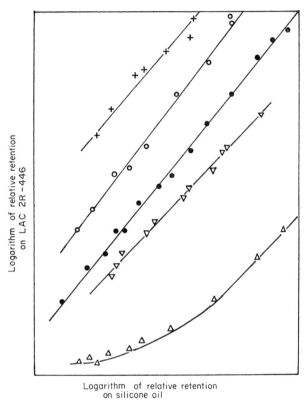

Fig. 51. Classification of organic compounds from retention data obtained from polar and non-polar columns. —×— aromatic ketones; —○— α,β-unsaturated ketones; —●— saturated ketones; —▽— arenes; —△— alkanes.

Retention data obtained using two or more different stationary phases is of more value. As shown in Fig. 51, the slopes of the plots of the retention data for homologous series measured on two columns are characteristic of the functional groups and thus provide a means of classifying a compound even although it may not be possible to definitely identify it. In many laboratories which use only two types of column, one being "polar" the other "non-polar" identification charts similar to that shown in Fig. 51 are readily constructed for the operational conditions which are standard for the laboratory. Unequivocal identification of an unknown component is possible by comparison of its retention data obtained using three stationary phases, one being non-polar and the other two being electron donating and electron accepting phases, with that of standard compounds[10] (see p. 61). It is extremely unlikely that two compounds which have the same degree of polarity will also have the same donor/acceptor properties.

The most accurate method by which the retention characteristics of a solute may be described is the *Kovats' retention index* system[11, 12]. This system, which is gaining in popularity particularly amongst organic chemists, is based upon a retention scale obtained from the logarithms of the adjusted retention volumes of the homologous series of n-alkanes. Not only can the system be used to identify compounds, it can also be used to predict the retention of solutes.

The retention index numbers for the n-alkanes are defined as $100\,n$ for the general formula C_nH_{2n+2}. Thus, the retention index for hexane is 600 and for heptane it is 700. As the logarithms of the adjusted retention volume of the n-alkanes increases linearly with the chain length, the arbitrarily chosen scale is linear. In the evaluation of retention indices with values of less than 100 the retention index of hydrogen is taken as zero. This corresponds to the dead volume of the instrument and, as indicated on p. 14, its measurement can present some difficulty depending upon the detector in use. In an isothermal measurement the retention index of an unknown compound may therefore be obtained by the interpolation of the logarithm of its retention volume between those of the relevant n-alkanes. This may be done graphically (see Fig. 52) or by use of the following equation:

$$I_X = 100 \cdot \frac{\log V_X - \log V_n}{\log V_{(n+1)} - \log V_n} + 100\,n$$

The temperature dependence of the retention index is small, so that the index can be used in a modified form for temperature programmed chromatograms (see p. 123). Although the temperature dependence is small, the structure of the compound and the type of stationary phase will have a considerable effect upon the value of the retention index. The retention indices of "non-polar" compounds remain almost constant for any type of stationary phase and their values depend largely upon their molecular shapes which govern their physical interaction with each

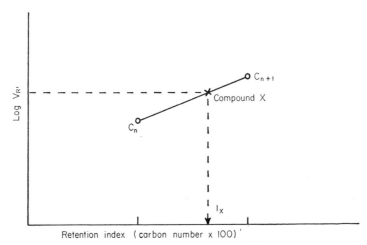

Fig. 52. Relationship between the logarithm of the adjusted retention volumes of n-alkanes and the Kovats retention index showing the method for the evaluation of the Kovats retention index of a component.

other and with other molecules. Generally the more linear is the molecule the higher is the value of its retention volume and consequently its retention index. Thus, the C_7H_{16} hydrocarbons, 2,2,3-trimethylbutane, 3,3-dimethylpentane, and 3-methylhexane have retention indices of 642, 662, and 676 respectively, when measured at 50°C irrespective of stationary phase. Also, the retention index of any compound, "polar" or "nonpolar", remains almost constant when measured on different "nonpolar" stationary phases. The gas chromatographic behaviour of a compound on a "polar" and a "non-polar" stationary phase, expressed in terms of its retention indices, may therefore be used to describe accurately the basic structure of the compound. The difference in the values of the retention index, ΔI, comprises additive terms for each skeletal and polar function of the structure[13].

$$\Delta I = I_{\text{polar}} - I_{\text{non-polar}}$$

From a knowledge of the retention index increments for the functional groups one is able therefore to predict a value for the retention index of a given compound or, alternatively, having measured the retention indices for an unknown compound on a "polar" and a "non-polar" column, one can assign a reasonable structure to the compound.

The following elementary examples illustrate the ways in which retention indices can be used. The retention index, measured on a "nonpolar" column, of a compound suspected of being a simple linear alcohol was found to be 1223. As this value is 200 units higher than that of

n-octanol measured under the same conditions, it may be inferred that the unknown compound was n-decanol. That the compound was indeed an alcohol was shown by measuring its retention index on a "polar" column. It was found that the ΔI value was identical to that obtained for n-alkanols measured under the same conditions. The full potential of the system is seen in the elucidation of the structure of a terpenoid alcohol. The ΔI value at 190° for the alcohol was found to be 270. Using the incremental values of ΔI at 190° for the cyclohexane ring, the alkene group and the hydroxyl group, together with incremental values for the alkyl substituents[13], ΔI values of 276, 281, 304 and 333 were calculated for the isomeric alcohols (I), (II), (III) and (IV). The unidentified alcohol was therefore considered to be most probably mentha-1-en-4-ol (I) and this was confirmed by direct comparison of its retention data with that of an authentic sample.

I   II   III   IV

2. FUNCTIONAL GROUP ANALYSIS

The structural assignments which have been made on the basis of retention data correlations may be corroborated by simple colour tests which are carried out directly on the effluent gas[14]. To obtain a measurable concentration of the component in the carrier gas, most probably, it will be necessary to overload the column. A stream splitter should be fitted to bypass the detector to allow the collection of the maximum amount of compound. The device shown in Fig. 53 enables several functional group tests to be made simultaneously. Of necessity, the reagents used for the colour tests must be extremely sensitive and a selection of simple colour tests which are suitable for the detection of amounts in the range 20–50 μg are given in Table V. More extensive lists of spot tests are available in the literature[14–16].

As an alternative procedure to aid the identification of unknown components, the entire mixture may be subjected to chemical reaction prior to gas chromatographic analysis. Thus, for example, unsaturated compounds may be catalytically hydrogenated using a platinum oxide catalyst and carbonyl compounds can be converted into hydroxy com-

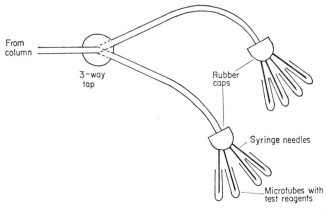

Fig. 53. Stream splitting device for the simultaneous testing of the column effluent with several reagents.

TABLE V

Functional Group Analysis by Colour Tests

Functional group	Reagent	Colour	Limiting sensitivity (μg)
Hydroxyl (alcohol)	$K_2Cr_2O_7/H^+$	Blue	20
Formyl	2,4-dinitrophenyl hydrazine	Yellow precipitate	20
Methyl ketones	2,4-dinitrophenyl hydrazine	Yellow precipitate	20
esters	$NH_2OH/FeCl_3$	Red/purple	40
Sulphides	Sodium nitroprusside	Red	50
Amines,			
primary	sodium nitroprusside	Red	50
secondary	sodium nitroprusside	Blue	50
Alkyl chlorides	alc. $AgNO_3$	White precipitate	20

pounds with sodium borohydride. Treatment with sodium removes all classes of compounds other than hydrocarbons and ethers. These reactions and others were originally employed with the sample in the vapour phase[17], but, with slight modification, they are equally suitable for pretreatment of the sample in the liquid phase.

3. THIN LAYER CHROMATOGRAPHY[18]

The technique of thin layer chromatography (TLC) is well known to be useful for the separation and identification of small concentrations of both volatile and involatile compounds. It is possible to correlate the retention of the TLC "spots", R_f, with those of standard compounds and identification of the functional groups can be made by spraying the plate with appropriate reagents, or alternatively, the "spots" may be removed from the plate and analysed spectroscopically[19, 20].

$$R_f = \frac{\text{distance of "spot" from starting point}}{\text{distance of solvent front from starting point}}$$

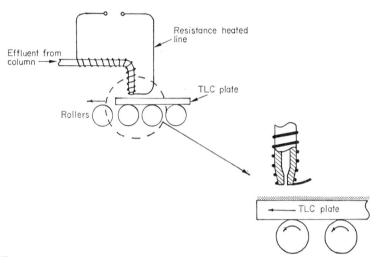

FIG. 54. Heated delivery capillary for transfer of gas chromatographic effluant to a thin-layer chromatographic plate.

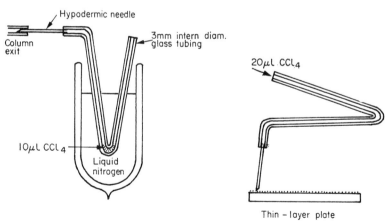

FIG. 55. Collection of micro samples by a microcrystalline porous frozen plug of solvent for subsequent transfer to a thin-layer chromatographic plate.

The obvious simplicity and cheapness of the system make the TLC–GLC combination a powerful procedure. The effluent gases are either absorbed directly on a TLC plate which is moved across the column exit such that the eluted components are deposited at intervals along the plate (Fig. 54) or, alternatively, the individual components are trapped in a small volume of solvent and subsequently transferred to the TLC plate as shown in Fig. 55.

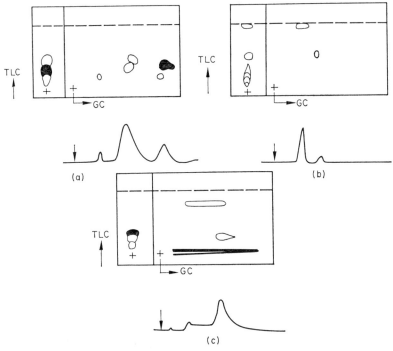

Fig. 56. Thin-layer chromatograms of GLC effluent showing a. insufficient separation on the gas chromatographic column, b. loss of part of the sample during gas chromatographic separation. c. decomposition of sample on the gas chromatographic column.

Not only can the TLC–GLC technique be used for the analysis of a mixture but it is also of considerable use in monitoring the efficiency of the gas chromatographic separation of the mixture. A sample of the total mixture is placed on the TLC plate together with the components obtained from the gas chromatographic column. Figure 56a illustrates the simple situation in which the components are inadequately separated by the gas chromatograph resulting in a non-homogeneous peak comprising two components. Figure 56b shows the situation in which the gas chromatographic separation is completely inadequate. Of the five components in the original mixture only two are eluted. Thin layer chromatography can also reveal the formation of artefacts and decomposition of components during gas chromatographic measurements (Fig. 56c).

4. PHYSICAL METHODS OF ANALYSIS

(a) *Introduction*: The most useful analytical methods which are generally available in the majority of laboratories are infrared (IR) and nuclear magnetic resonance (NMR) spectrometry. The former is particularly useful in the identification of functional groups but is of less value in distinguishing different members of a homologous series. NMR spectrometry is capable of defining the environment of every proton in the molecule and, with additional probes and the necessary ancillary equipment, the environments of other nuclei may also be investigated. Measurements of spectral absorption in the ultraviolet and visible region of the spectrum primarily provide information concerning the electronic structure. The molecules under investigation must contain chromophores and, in most cases, also auxochromes which will shift the absorption maxima to wavelengths above 200 nm such that they may be readily measured with the usual routine instruments. An unequivocal interpretation of the electronic absorption spectrum is normally not possible. Only Woodward's Rules provide a guide to structural assignments and are confined to unsaturated carbocyclic compounds. Mass spectrometers are not, in general, readily available, but for the structural analysis of samples obtained by GLC separation mass spectral measurements certainly have the greater potential. The molecular weight of the compound is obtained with little difficulty and, by a detailed analysis of fragmentation patterns, it is possible to reconstruct the full structure of the compound. Double focusing instruments are capable of giving accurate molecular weights which can often be correlated with an accurate molecular formula. Mass spectrometric analysis also has the advantage that it can be carried out directly on the sample retained in the carrier gas effluent whereas it is often necessary to collect the sample in milligram quantities in order to obtain useful data from the other spectroscopic methods.

The foregoing description serves only as an outline of the uses of the various spectroscopic methods of analysis and for those not aquainted with the techniques there are several introductory monographs which describe their theory and practice[21-24] and which also provide references to the more comprehensive texts which are of use in the interpretation of the data. The remaining sections of this chapter are concerned primarily with the various devices that can be used for the collection and manipulation of small quantities of the sample and in the design of micro apparatus for spectral analysis.

(b) Infrared Spectroscopic Analysis[21]: The most common of the analytical techniques for the identification of the separate components eluted from a gas chromatographic column is infrared spectroscopy. The main reason for its popularity results, in the main, from the relatively inexpensive nature of the auxiliary equipment compared with other analytical techniques and, to a lesser extent, the ease with which the relatively inexperienced analyst can interpret the spectra.

There are two main approaches by which the infrared spectrometer can be coupled to the gas chromatograph. The gas eluted from the chromatograph may be passed directly into an infrared gas cell and the spectrum of the component measured in the gas phase, or, alternatively, the component may be condensed and its spectrum measured either as a neat liquid or as a solution in a suitably infrared transparent solvent. The latter method is often preferred as it requires less sophisticated equipment. The major disadvantage of the method, however, resides in the isolation and manipulation of the sample. For solution spectra using standard size infrared cells, the minimum sample size required for the measurement of the medium to strong intensity absorption bands of a compound with a molecular weight of 200 is $c.$ 1 mg and, under normal conditions, it is not possible to obtain such a large amount from a standard chromatographic column. However, the introduction of microcells[22], which can have a capacity of as small as 0·5 μl, makes feasible the spectroscopic analysis of components separated on a standard packed column. The design of the microcell is similar to that of a standard size cell but the "dead volume" and cell aperture is reduced to a minimum such that almost the entire sample is confined in the area of the radiation beam. Cavity cells, i.e. cells produced by machining a slot cavity in a solid block of a suitable infrared optical material, are also commercially available. Different sizes of cavity are available giving path lengths from 0·01 mm upwards and with internal volumes as small as 0·2 μl.

In order to collect the components from the effluent gases, it is generally advisable to attach a stream splitter to the column outlet to divert the major proportion of the gas flow through a heated outlet to the cooled collection trap while a small fraction of the effluent gases flows through the detector. This modification to the outlet system is particularly important when a destructive detector, as for example the flame ionisation detector, is used but is not necessary for use with the non-destructive thermal conductivity detector. The efficient condensation of the components, however, presents a problem. Vigorous cooling is required to condense highly volatile compounds, whereas too great a temperature gradient results in the formation of fogs by the less volatile compounds which, instead of being trapped, are carried away by the carrier gas. In general, it appears that the induction of turbulant flow in the gas stream in contact with the cold surface improves the efficiency of condensation. Several collection systems are commercially available and various collection cells have been described[23]. The system shown in

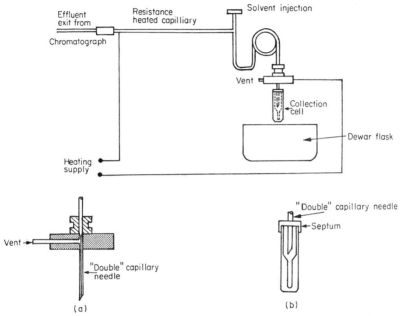

FIG. 57. Sample collection unit with a. heated "double" capillary needle b. glass trap.

Fig. 57 has been used with considerable success in the author's laboratory. The collection cell is essentially similar to that described by several other workers. The effluent gases enter the cell via the inner capillary of a "double" capillary needle and the component condenses at the neck of the reservoir while the carrier gas leaves by way of the outer concentric tube. The condensate is subsequently centrifuged into the capillary reservoir from which it can be conveniently removed with a microsyringe. This method of collection has the advantage that large amounts of sample can be accumulated by collecting fractions from several chromatographic separations. To obviate the somewhat difficult transfer of the sample from the trap to the infrared cell, the sample may be condensed, using the same collection system, directly in a silver chloride infrared microcavity cell. A suitable solvent is added from a microsyringe to the condensate, which is trapped in the upper part of the cell, and the solution is centrifuged into the optical region of the cell.

The infrared spectra of liquid samples have also been measured by incorporation of the compound in potassium bromide discs. Various methods have been described[23] by which potassium bromide is impregnated with the component directly from the effluent gas. A typical simple apparatus is shown in Fig. 58. For use in conjunction with gas chromatographs, ultramicro dies, which are capable of producing discs as small as

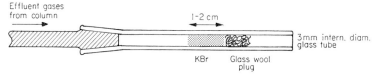

FIG. 58. Sample collection on KBr powder for subsequent IR analysis.

0·5 mm diameter requiring only c. 5 mg of potassium bromide and a sample size of the order of 2 µl or less, have recently become commercially available. In a similar technique the gas chromatographic sample is condensed in a millipore filter disc which is then examined directly in an infrared spectrometer[23]. This technique suffers however, from the disadvantage that the millipore material has strong spectral absorption regions.

Although these microtechniques are ideally suited for use with gas chromatographic separation, they produce an optical problem in that the apertures of the microcells and microdiscs are small and mask the radiation beam. This leads to a considerable loss of source energy. Some degree of compensation can be attained by equipping the spectrometer with an ordinate expansion unit which produces by electronic means up to a fivefold intensification of band intensities. The system is not entirely satisfactory, however, as sensitivity is lost and the background noise is also magnified. In an alternative and preferable technique the radiation beam is condensed and brought to focus as it passes through the infrared cell. In this manner a major proportion of the source energy passes through the cell aperture. The simplest commercially available beam condenser consists of two convex potassium bromide lenses, but more complex mirror systems are also available[24].

The infrared spectra of micro-samples have also been measured by the attenuated total reflectance technique[24].

Although measurement of infrared spectra in the gaseous phase requires more sophisticated apparatus than for liquid phase measurements, the technique has the advantage that there is no necessity to separate the sample from the carrier gas. Consequently "on-line" analysis is feasible and may be operated in two ways. The gaseous sample may either be isolated in the infrared cell during the measurement[25-28] or, alternatively, a continuous flow of the effluent gases through the cell is permitted[28, 29]. A typical arrangement whereby the sample is isolated during the spectroscopic analysis is shown in Fig. 59. Initially the

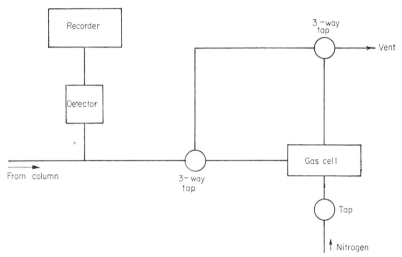

FIG. 59. Layout for through-flow infrared gas cell with a nitrogen flush system.

effluent gases are allowed to flow to waste until the recorder indicates the presence of a component. The flow is then directed through the gas cell. Several methods have been described by which it is possible to ascertain when the maximum concentration of component is contained within the cell. The simplest involves holding the spectrometer at a constant wavenumber setting at which it is known the component has an intense band. (This obviously requires a trial run prior to the analytical measurement.) When a satisfactory intensity is observed, the valves to the gas cell are closed and the full spectral range is scanned. After the measurement has been made, the cell is evacuated and flushed with nitrogen several times before the next sample is admitted. Thus, even when the spectroscopic analysis time is short, the complete procedure is time consuming, and without modification it is only suitable for the infrared measurement of selected components from a chromatographic separation. This difficulty of matching the speed of the chromatographic separation with that of the spectroscopic analysis has, however, been overcome by a technique described by Scott et al.[30] in which the carrier gas flow is interrupted during the spectroscopic analysis and only restarted after the cell has been flushed. The interrupted flow does not impair the efficiency of the chromatographic separation and the method is entirely satisfactory for the spectroscopic analysis of each component from a single chromatographic separation.

The effective usefulness of the alternative technique, whereby the effluent gases flow continuously through the cell, depends upon the scanning rate of the spectrometer. In principle, a spectrometer with a scanning time of the order of 5 seconds should be capable of recording a satisfactory spectrum of a 10 ml sample flowing through a 10 ml capacity gas cell at 1 ml/sec, but the many attempts to modify standard spectrometers have succeeded in producing shorter scanning times only at the expense of appreciable concomitant losses of sensitivity and resolution. The method is not generally very successful, but recently a repetitive scan system which utilises a Michelson interferometer coupled with a computer to store the spectral data has been described[29]. The system can not only compare the data with reference spectra, but it is also capable of resolving the spectra of components which are only partially resolved by the gas chromatograph.

The most important factors, however, in the gas phase analysis are the path length and volume of the gas cell. Depending upon the operation of the gas chromatograph, the total volume occupied by the component and the accompanying carrier gas is, in general, of the order of 10 to 20 ml, but the concentration of the component in this volume is low. Thus, a long path length gas cell (e.g. between 10 and 100 cm) is required in order to obtain a satisfactory spectrum of the sample. It is not usual for standard spectrometers to accommodate gas cells of path lengths greater than c. 20 cm and use is often made of a cell, which by the incorporation of mirrors to reflect the radiation beam several times through the gas, has an effective longer path length[24]. This type of cell has the added advantage that its total capacity can remain small while its effective pathlength is large. Such a cell, which also incorporates a beam condenser, is illustrated in Fig. 60. An alternative form of long path length gas cell, known as a "light pipe", does not utilise mirrors but depends upon the propagation of the radiation beam through the gas by multireflection off the highly polished internal surface of the cell[26, 27].

The capacity of the infrared cell should match, as near as possible, the volume occupied by the component and accompanying carrier gas. When the chromatographic band is broad such that the effective volume of the component is greater than that of the cell, only a fraction is accommodated in the cell and the sensitivity is appreciably diminished. Conversely, when several closely eluted narrow bands occupy the volume of the cell, the resolution obtained by the chromatograph is lost. It is usually difficult to alter the cell capacity and the incompatability between cell and sample volumes is more readily obviated by the adjustment of the chromatographic operational parameters.

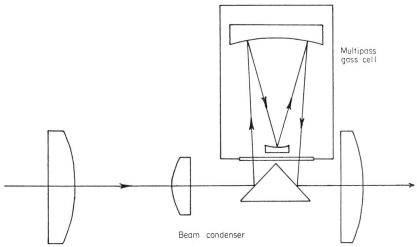

FIG. 60. A combined multi-pass micro gas cell and beam condenser.

(c) *Nuclear Magnetic Resonance*[31, 32]: Prior to the NMR measurement, the sample must be isolated from the effluent gases and the minimum quantity of sample required is usually considerably greater than that used for infrared analysis, but by repeated collection of fractions using the system described in the previous section (Fig. 56), it is possible to accumulate sufficient sample. Highly volatile compounds may be collected more conveniently by allowing the effluent gases to bubble through a small quantity of cooled carbon tetrachloride. In general, the minimum sample requirement for the resolution of a multiply split resonance signal from a single proton in a compound of molecular weight of 150–200, when measured using a standard sample tube, is between 1·0 and 10 mg and, for optimum sensitivity, the entire sample should be confined, as completely as possible, within the "active" volume of the receiver coil[33]. The "active" volume of the majority of coils has been shown to be between 0·2 and 0·25 ml, but it is preferable for routine work, or where optimum resolution is required, that a larger sample volume of *c.* 0·4 ml, which is more readily located in the probe, should be used. The sample requirement may be reduced by a factor of *c.* 5 when a microcell is used, and the increased signal to noise ratio available from instruments operating at higher field also permit the sample size to be reduced. The sensitivity of a 100 MHz spectrometer is approximately twice that of a 60 MHz instrument and it is feasible to obtain a satisfactory spectrum from c. 0·2 mg of sample in a 40 μl microcell with a 100 MHz spectrometer.

The shape of the sample volume and hence the design of the microcell is critical. In order to minimize flux distortion the sample shape should be preferably spherical. The performance of the cell, however, also depends largely upon the accuracy with which it can be positioned within the probe. An integral sphere microcell with a bulb capacity of c. 40 μl (Fig. 61a) and also a similar cell with a capacity of 100 μl (Fig. 61b) are commercially available. For routine analysis these cells are

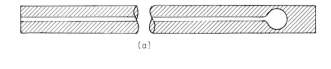

(a)

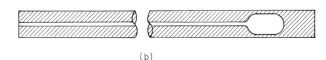

(b)

FIG. 61. All glass integral micro NMR cells a. c. 40 μl capacity b. 100 μl capacity.

satisfactory, but it has been reported that the performance with respect to resolution can be quite variable and that, in general, the larger volume cell is to be preferred. A significant cause for poor resolution, however, may often not be due to the microcell but to solid impurities in the sample. The "cleanliness" of the sample is considerably more important with microcells than with standard tubes, as the entire sample volume is held within the "active" volume of the receiver coils, and filtration of the sample through a porous glass disc immediately prior to its introduction into the microcell is recommended.

Several microcells which can be constructed quite readily in the laboratory have been described[31]. Figure 62 shows one such cell[34] which, in slightly differing forms, has attracted wide use. A bulb is blown in a capillary melting point tube until its external diameter is slightly less than the internal diameter of a standard NMR sample tube and its volume is approximately 30–40 μl. The bulb is completely filled with a solution of the sample (c. 0·2 to 0·5 mg) in carbon tetrachloride or deuterochloroform and the capillary is sealed. The microcell is supported in the NMR tube by a Teflon collar and the surrounding volume filled with carbon tetrachloride. The overall performance of this microcell is comparable with that of the commercial integral sphere microcell, but it cannot be used at elevated temperatures.

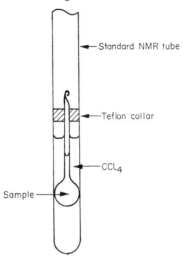

FIG. 62. Modification of a standard NMR tube for microsamples.

When the sample concentration is extremely small the signal to noise ratio is frequently low and can only be enhanced by data smoothing. A most powerful technique in this respect is that of time averaging which is a procedure whereby the signals from several spectral scans are combined to produce a composite spectrum[35]. As, in general, noise is a random process, it tends to cancel whereas the coherent resonance signals accumulate directly. Several electronic systems capable of time averaging electrical signals are commercially available. The unit, which is basically a digital memory in the form of a series of multichannel pulse counters, is generally referred to as a CAT ("Computer Average Transients"). The spectrometer repeatedly scans the resonance signals and feeds the signal at set frequency intervals to successive pulse counters. Finally the contents of the memory are converted back into an analogue voltage which actuates the recorder. Registration between the computer and spectrometer must be maintained and the spectrometer sweep is invariably controlled by a field frequency lock which ensures coherence between successive frequency sweeps. The resolution is limited almost entirely by the frequency intervals, a factor which depends upon the overall frequency sweep and the number of channels in the memory. For routine analysis, frequency intervals of c. 0·5 Hz using the commercially available 1024 channel system gives adequate resolution over a 500 Hz frequency sweep. When the complete spectrum has been established, the resolution of specific resonance signals can be enhanced by decreasing both the frequency sweep and the frequency intervals. The enhancement of the signal to noise ratio depends largely upon the residence time in each channel and increases approximately as the square root of the scan time. A satisfactory signal to noise ratio is usually obtained from sweep rates of 5 Hz/sec and at this rate a tenfold enhancement is attainable from c. 100 scans in 3 hours over a frequency sweep of 500 Hz. Figure 63 shows the improvement attainable for a 0·8 mg sample in an all glass microcell by time averaging 210 scans.

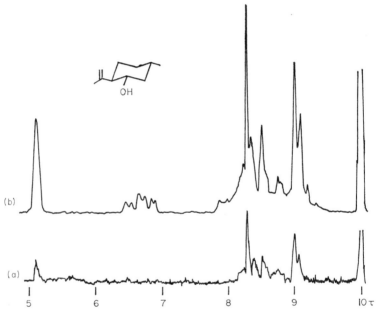

Fig. 63. NMR spectra of 0·8 mg sample of isopulegol in 40 μl CCl_4. a. Single scan at 60 MHz with scan rate of 1 Hz/sec. b. Time average of 210 scans at 60 MHz with scan rate of 2 Hz/sec.

(d) *Mass Spectrometric Analysis*[32, 36]: The combination of mass spectrometry and gas chromatography for the microanalysis of complex mixtures is the most effective technique available. Unfortunately, due in the main to the extremely high cost of the mass spectrometer, the technique is not generally available and its full potential has not been realised. The detection limits of the mass spectrometer are comparable with that of a flame ionisation detector although, in general, the sample requirement for a high resolution spectrum is of the order of 10 to 20 μg and a satisfactory mass spectrum can be obtained with a low resolution spectrometer from a 1 μg sample. The sensitivity of the mass spectrometer is such that too high a sample concentration is detrimental to the resolution and it is often necessary to split the chromatographic effluent stream and only allow a fraction into the spectrometer. The sample requirements are readily available from packed columns and, in favourable circumstances, open tubular columns are also capable of giving sufficient sample. A problem arises, however, from the use of a packed column in that "column bleeding" leads to a background spectrum of the stationary phase which invariably overlaps the spectrum of the sample. It is possible to avoid this type of contamination of the mass spectrometer by preferentially removing the eluted stationary phase with a short absorbant column which is inserted in the flow system immediately after the chromatographic column. Alternatively, the column bleeding may be minimised by conditioning the chromatographic column for several days at a temperature c. 40° higher than the operational temperature used in the analysis.

For routine analysis of organic compounds, the relatively inexpensive low resolution instruments are suitable. Two designs of instrument are available. The single focusing magnetic deflection mass spectrometer, which has an upper mass limit of 1500–2000 and a resolution of 1 unit in 1000, depends in its operation upon the separation by a magnetic field of the ions with a different mass to charge ratio (m/e). The second type of instrument, the time-of-flight spectrometer, is usually less expensive but has a somewhat lower resolution and mass range. The basis of its operation depends upon the different linear acceleration given to the ions of different m/e values under the influence of an electric field. The use of both types of instrument has been described for the analysis of gas chromatographic effluent gases[37–47]. The data from either mass spectrometer can be recorded by normal methods but, in order to exploit the extremely fast scanning rate, the use of a mirror galvanometer of small inertia working with UV light on UV sensitive paper is to be preferred. By this technique, mass peaks occupying no more than a millisecond can be recorded and the overall scanning time for 300–400 mass units at a resolution of 1 : 1000 is usually of the order of 2–3 seconds. With a lower resolution, the scanning time can be further reduced to c. 0·5 ~ 1·0 second. Thus, as both the sample requirement and the scanning rate of the mass spectrometer are compatible with the gas chromatograph, direct "on-line" operation is feasible and integral systems are commercially available. A limiting factor of the system, however, is the removal of the sample from the ionisation chamber. Some classes of compounds, particularly those with polar substituents or of high molecular weight, have a tendency to "stick" to the walls of the spectrometer. This consequently results in a high background of mass spectral peaks which overlap those of subsequently eluted compounds. The retained compounds are usually eluted under continuous pumping over a period of minutes, but occasionally a compound may have to be "baked out". Another source of background spectra results from poor chromatographic separation as a result of which the spectrum from the "tail" of an early component is superimposed upon the spectra of subsequent components. This problem is readily obviated by a better choice of chromatographic conditions. Two other important problems arise from direct coupling of the two instruments. A major requirement is a reduction in pressure from atmospheric in the gas chromatograph to the operational pressures of 10^{-4} to 10^{-6} torr in the ionisation chamber of the mass spectrometer. Most spectrometers are provided with a leak which controls the flow between the inlet system and the ionisation chamber, but it is usually necessary to establish a further pressure gradient by the use of a capillary link between the gas chromatograph and the mass spectrometer. When such a link is employed, it is imperative that it is maintained at a sufficiently high temperature to prevent condensation of the eluted components, otherwise the separation attained by the gas chromatograph is effectively reduced. Such a system, however, is required only for packed columns, as it has been shown that the necessary pressure gradient is established in the last few inches of a capillary column without detriment to the chromatographic separation. A direct link between the gas chromatograph and the mass spectrometer is therefore feasible, but such a system is not entirely satisfactory, as the whole of the effluent gases, with the carrier gas in excess, enter the ionisation chamber. The concentration of the sample is consequently often near or below the limiting sensitivity of the spectrometer. A further significant problem, encountered when the mass spectrometer is operated at the usual electron accelerating voltage of 70 eV, is that the carrier gas dominates the ion current and produces a high background on the total ionisation monitor. Ryhage[42] has overcome this problem by using helium as the carrier gas and operating the mass spectrometer at an electron accelerating voltage below the ionisation potential of helium (24·5 eV). Operation at low electron voltages, however, significantly alters the mass spectrum. In general, the parent peak is more pronounced and fewer

4. INTERPRETATION OF THE CHROMATOGRAM 111

fragmentation modes are detected. The alternative solution to the problem, and one which is now in general use, involves the concentration of the component by removal of the carrier gas. This procedure has the added advantage that it also effectively reduces the total pressure of the gas. Two forms of molecular separator are in general use. Both depend in their operation upon the difference between either the mass, or the volume, of the carrier gas molecules compared with those of the sample. Hence, for preference, helium is used as the carrier gas. The Watson-Biemann separator[45] is illustrated in Fig. 64. The effluent gases flow through a constriction which reduces the pressure to give a molecular

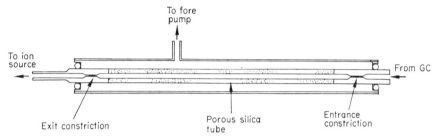

FIG. 64. A Watson-Biemann molecular separator.

flow within the porous tube. The volume outside the porous tube is continuously pumped and, as the rate of effusion of the lighter carrier gas is greater than that of the heavier sample, the concentration of the sample in the remaining gases is increased. The exit constriction provides a further pressure gradient to the ionisation chamber. One serious problem arises, however, from the use of the Watson-Biemann separator. Adsorption of the sample on the porous tube leads to tailing of the separated peaks with a consequent loss of resolution. At high temperatures the porous surface may also catalyse thermal decomposition of the components.

The less widely used Becker separator[48, 49] is shown in Fig. 65. A

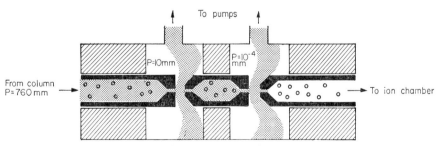

FIG. 65. A Becker molecular separator.

molecular flow of the effluent gases is caused to pass through a series of orifices while the volume about the orifices is continuously pumped. The heavier sample molecules maintain their flow but the lighter carrier gas molecules diffuse away from the flight path and are pumped away. The Becker separator is not suitable for use with capillary columns as the concentration of sample eventually reaching the ionisation chamber is below the acceptable level for a satisfactory mass spectrum.

A recent innovation in molecular separators is similar in concept to the Watson-Biemann separator but incorporates an organic semi-permeable membrane which permits the passage of the carrier gas and restricts the passage of the sample. A 0·005-in thick methyl silicone membrane enclosed in a glass envelope has been shown to effect an almost complete separation of the sample from the carrier gas with a 30% transfer of the sample to the mass spectrometer[50, 51]. The preliminary studies of this form of separator show no loss of resolution or thermal decomposition of the sample over a wide range of temperature and carrier gas flow rates and it is possible that the membrane separator will soon become the separator of choice.

Figure 66 illustrates a typical gas chromatograph-mass spectrometer assembly with a Watson-Biemann separator. The effluent gases from the column flow through a stream splitter, one fraction passing through the detector and a controlled fraction, the size of which depends upon the column size, enters the molecular separator. The flow of the sample through the ionisation chamber is recorded by a total ionisation monitor which collects a proportion of all ions prior to their deflection by the magnetic field. The monitor has the dual function of correlating the mass

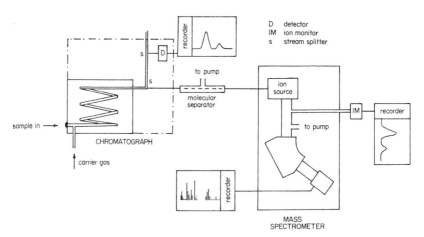

FIG. 66. Schematic diagram of the layout for a combined GLC-MS unit.

spectra with chromatographic peaks and also confirming that no change in the separation of the components has occurred during their flow through the connecting system.

Structural identification is best accomplished by comparison with mass spectra of standard compounds measured on the same instrument and under the same operational conditions. *A priori* identification of compounds from a correlation of the fragmentation pattern obtained at low resolution is possible but ambiguities often remain[52]. Such an analysis of complex molecules which contain elements in addition to C, H, or O, should be attempted preferably only with a high resolution instrument when the mass number of each fragment can be determined accurately to the third or fourth decimal place. A manual peak matching procedure is possible but alternatively the use of a computer which processes the data to produce an "elemental map" indicating the elemental composition of the parent ion and each fragment ion is more convenient. If sufficient prior information has been stored in the computer, it will also match the recorded spectrum with reference data and give the probable identity of the compound.

The high sensitivity of the mass spectrometer and the direct relationship between the intensity of a mass peak and the concentration of the

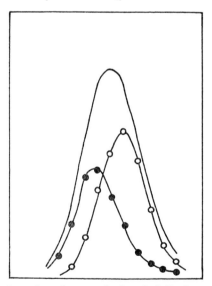

FIG. 67. Resolution of a mixture of trimethylsilyglucose and its heptadeutero form using a mass spectral comparison of the isotopic m/e 204 and 206 peaks (——— chromatographic peak on 3% silicone oil, —●— 206 mass spectral peak for the deutero compound —○— 204 mass spectral peak for the non-deutero compound).

fragment ion may be used to resolve completely both unresolved and partially resolved chromatographic peaks. The procedure depends upon the overlapping components having some distinct difference in their mass spectra. This difference need only be a mass difference in their most pronounced signals, as for example, the parent ions of a non-deuterated and deuterated compound. The mass spectral data is collected continuously as the unresolved chromatographic band flows through the ionisation chamber and is fed directly to a computer, which compares the relative intensities of the characteristic mass spectral signals of the two components and prints out their distribution in the chromatographic band (Fig. 67)[53]. Using a similar procedure, it is also possible to obtain the complete mass spectra of unresolved components.

As alternatives to the continuous "on line" analysis, the interrupted flow method described by Scott et al.[30] or a manual introduction procedure, whereby the component is initially isolated, may be used. For the latter technique, the component may be isolated by the method described on p. 101 and the sample subsequently transferred to the mass spectrometer as a liquid or solid. The sample, together with the carrier gas, may alternatively be trapped in a gas cell. The cell is subsequently cooled in an acetone : solid CO_2 bath or with liquid nitrogen and the carrier gas is removed under vacuum. The liquified sample is allowed to vaporize and is introduced directly from the cell into the mass spectrometer as a gas. A third method, which is not widely used, involves the collection of the component on potassium bromide or silica (cf. Fig. 58, p. 102) and the introduction of the component into the mass spectrometer as a "solid". Considerable care is required with these methods to exclude volatile contaminants and, for long term routine analyses, the manipulative procedures can be tedious. The method has the one advantage that any carrier gas may be used for the chromatographic separation.

References

1. Ongkiehong, L. *In* "Gas Chromatography, 1960" (R. P. W. Scott, ed.) Butterworths, London (1960).
2. Ronayheb, G. M., Perkins, G., Lively, L. D. and Hamilton, W. C. *In* "Gas Chromatography", Proceedings of the Third Symposium of the Instrument Society of America, 1961, (N. Brenner, J. E. Callen and M. D. Weiss, eds) Academic Press, New York (1962).
3. Ray, N. H., *J. Appl. Chem.*, **4**, 21 (1954).
4. Mefferd, R. B., Summers, R. M. and Clayton, J. D., *J. Chromatog.*, **35**, 469 (1968).
5. Bartlet, J. C. and Smith, D. M., *Can. J. Chem.*, **38**, 2057 (1960).
6. van der Vlies, C. and Caron, B. C., *J. Chromatog.*, **12**, 533 (1963).

7. Sawyer, D. T. and Barr, J. K., *Anal. Chem.*, **34**, 1213 (1962).
8. For a review of automatic digital integrators see Jones, H. J. *In* "The Practice of Gas Chromatography" (L. S. Ettre and A. Zlatkis, eds) Wiley, New York and London (1967).
9. Hawkes, S. J. and Russell, C. P., *J. Gas. Chromatog.*, **2**, 186 (1964).
10. Brown, I., *J. Chromatog.*, **10**, 284 (1963).
11. Kovats, E., *Helv. Chim. Acta.*, **41**, 1915 (1958).
12. For a review of the system see Kovats, E. *In* "Advances in Chromatography", (J. C. Giddings and R. A. Keller, eds) Vol. 1, p. 229. Edward Arnold, London (1965).
13. Wehrli, A. and Kovats, E., *Helv. Chim. Acta*, **42**, 2709 (1959).
14. Walsh, J. T. and Merritt, C., *Anal. Chem.*, **32**, 1378 (1960).
15. Feigl, F., "Spot Tests in Organic Analysis", 6th edition, Van Nostrand, Princeton (1960).
16. Schneider, F. L., "Qualitative Organic Microanalysis" Academic Press, New York (1964).
17. Hoff, J. E. and Feit, E. D., *Anal. Chem.*, **35**, 1298 (1963); **36**, 1002 (1964).
18. Kaiser, R., *Chemistry in Britain*, p. 54, (1969).
19. Bobbitt, J. M., "Thin Layer Chromatography", Reinhold, New York (1963).
20. "Thin Layer Chromatography", (E. Stahl, ed.) Academic Press, New York 1965).
21. Littlewood, A. B., *Chromatographia*, **1**, 223 (1968).
22. see, e.g., Haslam, J., Jeffs, A. R. and Willis, H. A., *Analyst*, **86**, 44 (1961).
23. Thomas, P. J. and Dwyer, J. L., *J. Chromatog.*, **13**, 366 (1964).
24. Martin, A. E., "Infrared Instrumentation and Techniques", Elsevier, Amsterdam (1966).
25. Flett, M. St. C. and Hughes, J., *J. Chromatog.*, **11**, 434 (1963).
26. Bartz, A. M. and Ruhl, H. D., *Anal. Chem.*, **36**, 1892 (1964).
27. Wilks, P. A. and Brown, R. A., *Anal. Chem.*, **36**, 1896 (1964).
28. White, J. U., Alport, N. L. and Ward, W. M., *Anal. Chem.*, **31**, 1267 (1959).
29. Low, M. J. D. and Freeman, S. K., *Anal. Chem.*, **39**, 194 (1967).
30. Scott, R. P. W., Fowlis, I. A., Welti, D. and Wilkins, T. *In* "Gas Chromatography 1966" (A. B. Littlewood ed.) p. 318, Institute of Petroleum, London (1967).
31. Lundin, R. E., Elsken, R. H., Flath, R. A. and Teranishi, R., *Appl. Spectr. Rev.*, **1**, 131 (1967).
32. Teranishi, R., Lundin, R. E., McFadden, W. H. and Scherer, J. R. *In* "The Practice of Gas Chromatography" (L. S. Ettre and A. Zlatkis, eds) p. 438, Interscience, New York (1967).
33. Nelson, F. A. *In* "NMR and EPR Spectroscopy" Pergamon, London (1960).
34. Flath, R. A., Henderson, N., Lundin, R. E. and Teranishi, R., *J. Appl. Spectry.*, **21**, 183 (1967).
35. Lundin, R. E., Elsken, R. H., Flath, R. A., Henderson, N., Mon, T. R. and Teranishi, R., *Anal. Chem.*, **38**, 291 (1966).
36. Littlewood, A. B., *Chromatographia*, **1**, 37 (1968).
37. Gohlke, R. S., *Anal. Chem.*, **31**, 535 (1959).
38. Ebert, A. A., *Anal. Chem.*, **33**, 1865 (1961).
39. Miller, D. O., *Anal. Chem.*, **35**, 2033 (1963).
40. Lindeman, L. P. and Annis, J. L., *Anal. Chem.*, **32**, 1742 (1960).
41. Holmes, J. C. and Morrell, F. A., *Appl. Spectry.*, **11**, 86 (1957).

42. Ryhage, R., *Anal. Chem.*, **36**, 759 (1964).
43. McFadden, W. H. and Day, E. A., *Anal. Chem.*, **36**, 2362 (1964).
44. Stenhagen, E. *Z. Anal. Chem.*, **205**, 109 (1964).
45. Watson, J. T. and Biemann, K., *Anal. Chem.*, **36**, 1135 (1964); **37**, 844 (1965).
46. Dorsey, J. A., Hunt, R. H. and O'Neal, M. J., *Anal. Chem.*, **35**, 511 (1963).
47. Biemann, K. and Watson, J. T., *Monatsh.*, **96**, 305 (1965).
48. Becker, E. W., "Separation of Isotopes", (H. London ed.) p. 360, Newnes, London (1961).
49. Ryhage, R., *Anal. Chem.*, **36**, 759 (1964); *Arkiv Kemi*, **26**, 305 (1967).
50. Black, D. R., Flath, R. A. and Teranishi, R., *J. Chrom. Sci.*, **7**, 284 (1969).
51. Hawes, J. E., Mallaby, R. and Williams, V. P., *J. Chrom. Sci.*, **7**, 690 (1969).
52. see e.g. Hill, H. C., Reed, R. I. and Robert-Lopes, M. T., *J. Chem. Soc., (Section C)*, p. 93 (1968).
53. Sweeley, C. C., Elliot, W. H., Fries, I. and Ryhage, R., *Anal. Chem.*, **38**, 1549 (1966).

Chapter 5

CHROMATOGRAPHIC TECHNIQUES

A.	Temperature Programmed Gas Chromatography	117
	1. Introduction	117
	2. Variable Parameters	120
	3. Retention Temperature	122
B.	Cryogenic Gas Chromatography	123
C.	Flow Programmed Gas Chromatography	124
D.	Backflushing	124
E.	Automated Preparative Chromatography	127
	1. Preparative Columns	127
	2. Sample Introduction	128
	3. Column Operational Cycle	129
	4. Traps	130
F.	Automatic Routine Analysis and Product Control	130
References		132

A. Temperature Programmed Gas Chromatography[1]

1. INTRODUCTION

If the boiling points of the components of a mixture are considerably different, it will be found that at no one temperature are the conditions optimum for their analysis on a non-polar column. At low temperatures, which permit satisfactory resolution of the low boiling fraction, the bands of the higher boiling components are broadened by diffusion. This leads to a loss of detector response particularly for small samples such that quantitative measurements will be inaccurate. Alternatively, when the column temperature is sufficiently high to elute the high boiling components efficiently, the more volatile components are eluted too rapidly with concomitant loss of resolution. Ideally, fractional distillation of the sample followed by chromatography of the higher and lower boiling fractions at different column temperatures would solve the problem, but this is not always possible. Similar problems arise in the isothermal analysis of components of widely differing polarity when chromatographed on polar columns.

In programmed temperature gas chromatography the column temperature is increased at a predetermined rate during the chromatographic measurements such that it permits the less strongly retained components to be resolved at a low temperature whilst the late components are eluted as sharp peaks at a higher temperature (Fig. 68a). As the temperature is increased the velocity of the components increases such that the velocity is approximately doubled by a 30° increase in temperature. It follows that under linear temperature programming conditions the individual components arrive at the middle of the column when the column temperature is approximately 30° below that at which the components are eluted. Thus, the final distribution ratio and consequently the peak widths are similar for all components. The heating

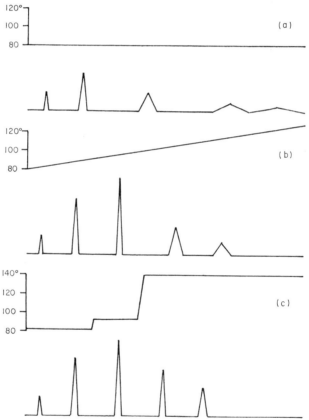

Fig. 68. Effect of temperature programming upon retention volume. a. Isothermal operation b. Linear temperature programme. c. Stepped temperature programme.

rate can be optimised so that each peak gives the appearance of having been chromatographed isothermally at the optimum column temperature.

The technique also has the advantage that, in most instances, the overall analysis time is less than that under isothermal conditions.

The increase in column temperature is usually linear with time (Fig. 68b). The heating rate can be varied and is generally between 0·5 and 50°C/min depending upon the initial and final temperatures and the analysis time for optimum separation. Alternatively, the temperature programme may consist of a series of isothermal steps with a rapid increase in temperature at the transition points (Fig. 68c). The intrinsic efficiency of the column will remain almost constant for isothermal operation, for linear temperature programming, and for stepped temperature programming. Any loss in column efficiency resulting from the temperature programme is due to non-uniform distribution of the components on the column. For this reason, the preferential use of capillary columns with the technique is to be recommended, although, in general, satisfactory results may be obtained with narrow diameter packed columns. A significant disadvantage of temperature programming is that the rate of "bleed" of the stationary phase from the column is not constant but increases logarithmically with the temperature. This results in a baseline drift which steadily increases with the increase in temperature. The effect is so great that with single column operation it becomes impossible to detect any peaks at high temperatures (Fig. 69a). To overcome the effects of the base line drift a second column, having as near identical characteristics as possible to those of the analytical column, is mounted

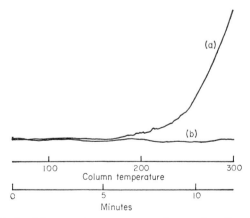

FIG. 69. Effect of temperature programming on baseline as result of column bleeding. a. Single column operation b. Dual column operation.

in the oven and connected to the reference detector. The vapour pressure of the stationary phase in each column will be almost identical so that if the signal from the reference detector is given the opposite polarity to that of the analytical detector and the two signals are fed simultaneously to the recorder then the base line drift is compensated (Fig. 69b). As it is difficult to prepare two completely identical columns, a slight base line drift will still be observed. This may be corrected by adjusting the carrier gas flow through the reference column. The flow rates through the two columns are initially made equal and the oven temperature is increased to the upper limit of the programme. The drift in the base line is then corrected by changing the flow rate through the reference column. The oven temperature is returned to the initial setting and the base line adjusted to zero by the pen control.

This procedure does not of course prevent column bleeding and, although satisfactory for most purposes, it is inadequate for use with detectors which have a high sensitivity. Also, collected samples will be contaminated and the usefulness of direct "on-line" mass spectrometry will be marred by background mass spectra of the stationary phase. A possible remedy has recently been reported[2] in which the column "bleed" is absorbed by a short column containing a stationary phase of low volatility.

2. VARIABLE PARAMETERS

The retention time of the solute is dependent upon the initial column temperature, the programmed heating rate, and the carrier gas flow rate. Thus, as for isothermal chromatography, a compromise often has to be reached in the selection of these parameters to obtain the optimum operational conditions. The choice of the initial temperature is virtually predetermined by the most volatile components of the sample, as the temperature should be such that the peaks are well resolved without the retention times being excessively long. This temperature often corresponds closely to the boiling point of the most volatile component. Also as a rough guide, it is often found that the most satisfactory initial temperature results in the retention volume of the first peak being approximately four times that of the air peak. If retention temperatures are used to characterise the components (*vide infra*) a limitation is imposed upon the upper limit of the initial temperature. It has been found that under a constant temperature programming rate the temperature at which a component is eluted normally remains constant over a range of initial column temperatures. However, above the upper limit of this range the elution temperature increases as the initial column temperature is increased. The upper limit of the range varies with the volatility of the component (Fig. 70) being lower for the more volatile compounds. Thus, if the retention temperatures of the components are to be maintained constant, relative to one another, the initial column temperature must be lower than the upper limit of the linear range for the most volatile component. If this limit is exceeded it is quite possible for the order of elution of the components to be altered.

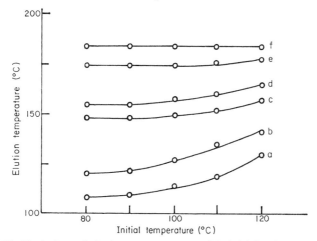

FIG. 70. Variation of elution temperature with initial column temperature. *a* hexanol; *b* decane; *c* and *d* dodecane isomers; *e* tetradecane; *f* dodecanol.

It is equally important that the rate at which the column temperature is increased during the chromatographic measurement should be such that it gives optimum resolution and minimum analysis time. The general effect of the heating rate is analogous to that of the column temperature in isothermal operation, i.e. both the resolution and the retention time increases with a decrease in the heating rate. Thus a low rate of heating should be used when possible and the resulting increase in the analysis time should be compensated by increasing the carrier gas flow rate. High carrier gas flow rates, however, reduce the column efficiency and a compromise in the choice of flow and heating rates has to be established for optimum operational conditions. Also, as the temperature of the column increases, the increased resistance of the liquid phase to the flow of the carrier gas produces a decrease in the flow rate. This is of little importance if the chromatograph is equipped with a mass detector, which is insensitive to the flow rate (see p. 77), but it is imperative for quantitative measurements that all instruments fitted with a concentration detector, such as the thermal conductivity detector, should be equipped with a flow regulator which maintains a constant carrier gas flow rate at all temperatures.

The upper temperature limit for programmed chromatography is determined, in the main, by the stability of the stationary phase (see p. 153).

3. RETENTION TEMPERATURE

As indicated in the preceding section, retention times and volumes are extremely sensitive to changes in the operational conditions. The general effect of the changes in the carrier gas flow rate is the same for temperature programmed operation as it is for isothermal chromatography (see p. 65) and the effect upon the retention times resulting from changes in the initial column temperature and the programmed heating rate are shown in Fig. 71. It is significant that the retention times of most solutes become almost constant when the programmed heating rate is

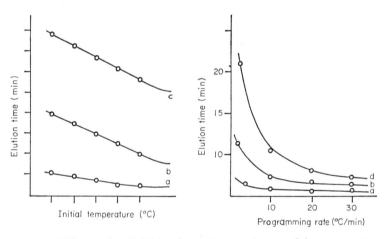

Fig. 71. Effects of a. initial column temperature and b. temperature programming rate upon sample retention. *a* hexanol; *b* dodecane; *c* tetradecanol; *d* hexadecanol.

high (above *c.* 20°C/min) and consequently there is very little saving in analysis time by the use of extremely high heating rates.

For temperature programmed gas chromatography, a considerably more reliable parameter which is characteristic for each solute/stationary phase system is the *retention* (or *elution*) *temperature*. This is defined as the column temperature at which the centre of the solute peak is eluted from the column and may be calculated from the following

equation
$$T_R = T_I + n.x.u$$

where T_R and T_I are the retention temperature of the solute and the initial column temperature respectively, n is the chart distance, measured in inches, between the injection point and the mid-point of the peak, x is the chart speed (inches/min) and u is the programmed heating rate given in °C/min. The relationship assumes that the heating rate is linear and that it commenced from the time of injection.

The retention temperature is virtually constant for all values of the initial column temperature below a critical upper limit (*vide supra*). Also, as it varies as a logarithmic function of the heating rate and carrier gas flow rate, it is relatively insensitive to small fluctuations in operating conditions. A direct correlation can be made between the retention temperature of a solute measured under linear temperature programmed conditions and the logarithm of the retention volume of the solute as measured under isothermal conditions. Thus, there is a linear increase in retention temperature with carbon number for a homologous series and it follows that there is also a direct correlation between the boiling point of a solute and its retention temperature. In the same manner that one can obtain the Kovats' retention index from log V_R measurements obtained from isothermal operation, so the direct use of the retention temperature will give the retention index for programmed temperature chromatography.

$$I = 100 \frac{T_{R(\text{solute})} - T_{R(n)}}{T_{R(n+1)} - T_{R(n)}} + 100\,n$$

where $T_{R(\text{solute})}$, $T_{R(n)}$, $T_{R(n+1)}$ are the retention temperatures of the solute and normal paraffins having n and $n+1$ carbon atoms[3].

B. Cryogenic Gas Chromatography

Just as it is important to establish optimum column temperatures for the separation of compounds having high boiling points, it is obvious that sub-ambient temperatures will improve the resolution of gases and liquids having low boiling points. Cryogenic gas chromatography has been found to be of use in the analysis of food volatiles and of refinery gases[4, 5] but the technique has not, as yet, been widely used, primarily due to difficulties encountered in maintaining constant subambient temperatures. Recent developments[6, 7], however, have produced reliable control systems which allow both isothermal operation and controlled temperature programmes over a continuous range of temperatures from -195°C to $+500$°C and several commercial units are available.

C. Flow Programmed Gas Chromatography[8]

As an increase in the carrier gas flow rate produces a decrease in the retention time of the sample, flow programming can be used in much the same way as temperature programming. The average flow rate through the column varies linearly with the pressure drop across the column. Thus, to reproduce the effect of linear temperature programmed gas chromatography, it is necessary to increase the column inlet gas pressure expotentially with time. Flow programming, however, has several important advantages compared with temperature programming. With the sophisticated control systems which are now available it is often more convenient to measure the carrier gas flow rate and to control the inlet pressures than to control the column temperature. Also, as a lower column temperature can be used in conjunction with flow programming, it is a more suitable technique for the analysis of thermally labile compounds. The lower column temperature also results in a significantly smaller loss of stationary phase by column bleeding, thus extending the life of the column and improving the detector performance. The time saved in the rapid re-establishment of initial operational conditions, compared with the long cooling down period necessary between analyses carried out under temperature programming conditions, can be of considerable importance to laboratories involved in an extensive programme of routine analyses. On the other hand, however, high carrier gas flow rates result in a loss of column efficiency. This is particularly noticeable with the appearance of the late peaks, the resolution of which is usually considerably better with temperature programming than with flow programming techniques. The flow programming procedure should also be used only in conjunction with detectors which are insensitive to the carrier gas flow rate. The most suitable is the flame ionisation detector, although even with this detector the change in ratio of the carrier gas to hydrogen may alter the response slightly.

As with temperature programming procedures, stepwise flow programming has been used and analyses using combined temperature and flow programming have been described[9] and it appears that a combination of the techniques is particularly successful when used with open tubular columns. Although only a recent innovation for routine gas chromatographic analysis, several linear and expotential inlet pressure programming units are already commercially available.

D. Backflushing

It is often found to be desirable to remove components from a column before they reach the detector. Such is the case, for example, when the mixture contains components with inordinately large retention volumes. These components may be of little interest and their elution under normal conditions would make the analysis time prohibitively long. Alternatively, although the less strongly retarded components are satisfactorily separated, the analyses of the remaining components may be more readily effected on a different stationary phase and their removal from the column would be advantageous.

Using the technique known as *backflushing*, the direction of the flow of the carrier gas through the column is reversed after the components of analytical interest have been eluted and analysed. In this manner the retained components are swept back through the column, regrouped, and eluted as a single band. This simple procedure has an obvious use in clearing the column of relatively involatile solutes which may interfere with later analyses.

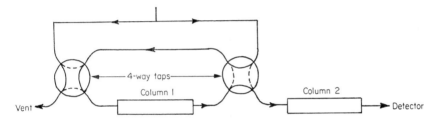

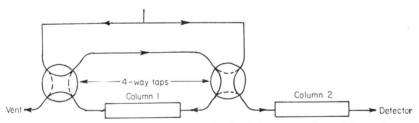

FIG. 72. Flow schematic diagram of a backflushing unit. a. Through-flow on both columns. b. Reverse flow through one column.

A more elaborate system is illustrated in Fig. 72. The normally eluted components are transferred to a second column and the retained components may be backflushed either to a third column for further analysis or to a vent. The pressure at the inlet to the second column is maintained at a constant value irrespective of the direction of flow through the first column. There are several variations of this technique. The regrouping procedures, in particular the so called "flip-flop" technique[10, 11], are often preferred in process control analysis where speed and reliability are important.

On account of their finite volume, the use of the four-way valves, as shown in Fig. 72, can reduce the efficiency of separation and the alternative use of remote pressure control valves, which are not in the sample path, have been recommended[12]. A unit employing the remote valves is illustrated in Fig. 73. Although the principle of the system is similar, it differs from that used for backflushing in that it is designed to

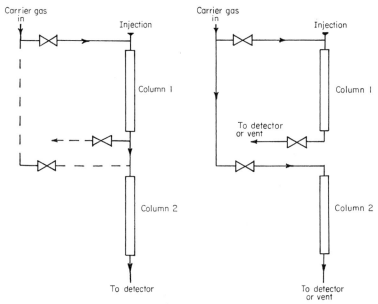

Fig. 73. Flow schematic diagram for fraction cutting by a split flow technique.

remove a single component or a group of components from any section of the chromatogram. The procedure is of particular use, for example, to remove a component of high concentration which masks trace components having slightly longer retention times. The technique can also be used to eliminate the solvent peak from the chromatogram of dilute solutions or to obtain a better resolution of selected components by cutting them from the remainder of the sample.

The flow of the carrier gas is initially continuous through both columns, i.e. valve A is open and valves B and C are closed. In order to separate the selected components, valves B and C are opened. The earlier components continue to be separated on column 2 whilst the selected components are eluted from column 1 and may be carried to a third column for further analysis, or alternatively eluted to the atmosphere. The system may then be returned to normal flow conditions for the analysis of the remaining components by closing valves B and C.

E. Automated Preparative Chromatography

I. PREPARATIVE COLUMNS

In a previous chapter (p. 100) several methods suitable for the isolation of microquantities of components from the column effluent were described. The techniques enable the analyst to obtain microgram samples for routine spectroscopic analysis but for larger quantities the routine of repeated manual injection and isolation becomes extremely tedious. When a standard analytical packed column is used, the major limiting factor, which controls the number of repetitive separations required for the accumulation of gram quantities, is the size of the sample which can be injected onto the column without imposing severe overloading of the stationary phase. The larger the column diameter, the greater is the capacity of the stationary phase and, in general, columns for preparative separation have internal diameters of between ⅜ in and 1 in (10 to 25 mm) although frequent use is made of macroscale preparative columns of up to 4 in (100 mm). Obviously, with standard chromatographs the oven

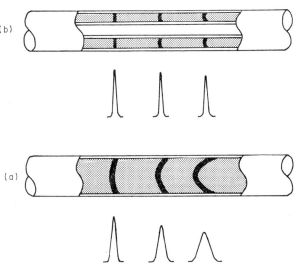

Fig. 74. Comparison of peak profiles of components eluted from a. a conventional wide bore column, b. a "biwall" column.

size then becomes an important factor and it is often the case with many analytical and routine instruments that increased column diameter must of necessity result in considerably shorter column length and a consequent loss in resolution (see p. 12). Also, as a consequence of the wide diameter, the sample vapour band does not flow in a uniform manner. Due in part to a decrease in packing density, the vapour velocity is greater in the vicinity of the column walls than at the centre of the stationary phase (Fig. 74a) and this effect often produces overlapping of components. The velocity gradient, which is only significant with wide diameter tubes, may be obviated by an annular "biwall" construction in which the stationary phase is supported between two concentric tubes (Fig. 74b). As the volume of the stationary phase per unit length of column is only slightly reduced, there is little alteration in the permitted sample loading but the narrower cross section of the stationary phase results in a considerable improvement in resolution.

2. SAMPLE INTRODUCTION

Automatic injection of the sample from a reservoir onto the column may be accomplished by means of a pneumatic system or by a mechanical pumping system and its operation is synchronous with the column operational cycle (*vide infra*).

A typical pneumatic injection system is illustrated in Fig. 75. Its

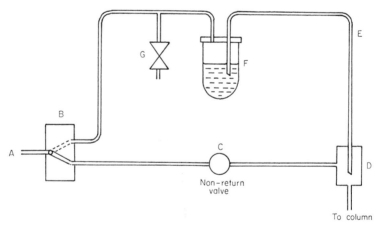

FIG. 75. An automatic pneumatic injection system.

A—carrier gas inlet   B—solenoid switching valve
C—non-return valve   D—injection port
E—capillary   F—reservoir
          G—needle inlet valve

operation involves the displacement by the carrier gas of a quantity of sample from the reservoir and its transfer to the column without contact with any valves. The carrier gas normally flows from A to D and thence through the column. From D it also flows, via the capillary E, through the sample in the reservoir F. In order to prevent any flow of sample to the column, the gas pressure above the sample in the reservoir is maintained at a value slightly below that of the column inlet pressure by a needle valve G. The programmer which controls the column operational cycle also triggers the solenoid valve B such that the carrier gas flow from B to D is blocked and the pressure above the sample in the reservoir increases simultaneously with a decrease in pressure at D. The sample is consequently forced through the capillary E onto the column. When the solenoid valve reverts to its former position the pressure is re-established at D and the sample is cleared from the capillary by a flow of gas towards the reservoir. A sharp injection of the sample is thus ensured and, depending upon the time cycle of the solenoid, precise quantities over a range of c. 10 μl to 5 ml can be introduced onto the column.

3. COLUMN OPERATIONAL CYCLE

Automatic preparative separation of components from a mixture consists of four discrete operations: (a) injection of the sample; (b) isothermal or programmed temperature separation of the components with the

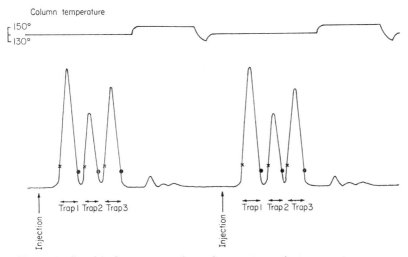

FIG. 76. Graphical representation of an automatic temperature programmed chromatographic cycle (—×— collection traps open, —●— collection traps closed).

subsequent collection of the individual components in separate traps; (c) increase in column temperature to remove residual components rapidly and (d) decrease in column temperature to the operational temperature for further injection and separation. The sequence of events which are illustrated graphically in Fig. 76, are controlled electronically. The valves for the individual traps are triggered to open and close in a prearranged sequence at pre-set detector voltages such that the components are delivered to their own individual traps during each cycle.

4. TRAPS

The conditions which are favourable for the condensation from the effluent gas of components on the microscale (see p. 100) are applicable to macroscale separation. The shape of the trap is therefore important and a selection of typical traps, which are commercially available, are shown in Fig. 77.

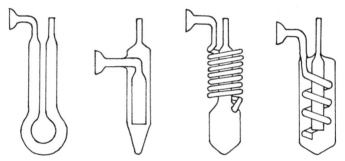

FIG. 77. A selection of traps suitable for a wide range of sample types.

F. Automatic Routine Analysis and Product Control

Industrial product control analysis by gas chromatography, often called *process gas chromatography*, requires automatic, repetitive sample introduction and an easy-to-read presentation of the data.

For routine batch analysis of solid materials, several automatic injection systems, which can be used with standard analytical gas chromatographs, have been developed. The introduction of the sample, retained in glass capillaries or impregnated on gauze pellets, is controlled by a programmer unit which operates in a similar manner to that described for automatic preparative gas chromatography. The systems may also be used for liquids if their volatility is such that pre-vaporisation of the sample does not occur prior to its introduction into the flash heater. An automatic system, which can be used for both solid and liquid samples, in which the samples are introduced from sample bottles onto the column by means of a standard syringe and similar sampling technique for the analysis of head space gases are also commercially available. The automatic injection system, which is shown in Fig. 78, utilises glass capillary sample holders.

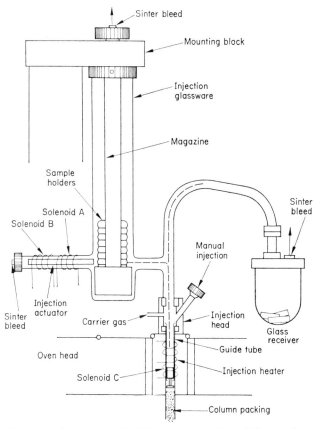

FIG. 78. An automatic injection unit for solid samples.

For on-line product control analysis, the sample-handling procedure involves continuous sampling from the production stream and, in order that the chromatograph can analyse the sample efficiently and accurately, it is necessary to "clean" and "regulate" the sample. The sample "conditioner", which for obvious reasons must be specifically designed for the particular process, consists of a pressure regulator and filters and traps to remove contaminants. If liquid samples are to be analysed, a vaporiser unit must also be interposed between the production line and the chromatograph. The sampling system shown in Fig. 79 is a simple unit for gas analysis.

Since on-line process analysis requires the installation of the gas chromatograph in the operational plant, it must be more robust than its laboratory counterpart and its operation is subject to rigid safety regulations. Not only must it be capable of withstanding extreme operational conditions, often in corrosive or inflammable atmospheres, but it must also provide long service with the minimum of maintenance.

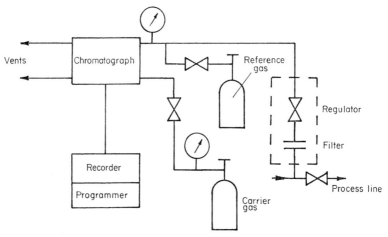

FIG. 79. Schematic diagram of an industrial on-line gas-liquid chromatograph.

Data presentation for both batch analysis and process analysis is normally in the form of digital print-out, but there is an increased tendency towards the transmission of analytical data from process monitors directly to a computer which can automatically control production.

References

1. Harris, W. E. and Habgood, H. W., "Programmed Temperature Gas Chromatography", Wiley, New York (1966).
2. Levy, R. L., Gesser, H. D. and Hougen, F. W., *J. Chromatog.*, **30**, 198 (1967).
3. Van den Dool, H. and Kratz, P. D., *J. Chromatog.*, **11**, 463 (1963).
4. Merritt, C. and Walsh, J. T., *Anal Chem.*, **35**, 110 (1963).
5. Jones, C. N., *Anal. Chem.*, **39**, 1858 (1967).
6. Ehrler, A. J. and Maczka, H., Victoreen Research Report. Victoreen Instrument Division, Cleveland, Ohio. (based on paper presented by A. J. Ehrler at the Conference on Analytical Chemistry and Applied Spectroscopy, Cleveland, Ohio, 1968).

7. Robertson, D. H., Issenberg, P. and Merritt, C., "Facts and Methods", Vol. 5, Hewlett-Packard Avondale, Pennsylvania. (1964).
8. Scott, R. P. W. *In* "Progress in Gas Chromatography" (H. Purnell, ed.) Interscience, New York (1968).
9. Mazor, L. and Takacs, J., *J. Chromatog.*, **29**, 24 (1967).
10. Villalobos, R., Brace, R. O. and Johns, T., *In* "Gas Chromatography" (H. J. Noebels, R. F. Wall and N. Brenner, eds) Academic Press, New York (1961).
11. Littlewood, A. B., *J. Gas Chromatog.*, **2**, 188 (1964).
12. Deans, D. R., *J. Chromatog.*, **18**, 477 (1965).

Part II
Appendix

A. Glossary of Gas Chromatographic Terms, Symbols and Abbreviations

This glossary comprises the definitions of terms, symbols and abbreviations used in this book. The recommendations of the IUPAC committee for gas chromatographic nomenclature have been observed wherever possible. Other definitions, not covered by the committee's recommendations, are those which are in current general use.

air peak The peak of a substance which is not retarded by the stationary phase (e.g. air or an inert permanent gas). The retention volume corresponding to the peak is equivalent to the dead volume of the instrument.

attenuator A stepped resistance control by which the input voltage to the recorder may be altered.

backflushing A technique in which the direction of the carrier gas flow through the column is reversed in order to remove involatile solutes.

band see *peak*.

base line The recorded detector response when only carrier gas is passing through the detector.

carrier gas The inert gas which acts as the mobile gas phase in gas chromatographic separations.

chromatogram A plot of the detector response against time or the volume of carrier gas.

column A tube containing the stationary phase. In *packed columns* the stationary phase is supported on an inert solid. In *open tubular* or *capillary columns* the stationary phase is coated as a thin film on the internal surface of the tube.

column "bleeding" The elution of the liquid phase from the solid support by the flow of the carrier gas. The rate of elution of the liquid phase increases with increase in temperature.

column, ideal and non-ideal The input distribution of the solute is not broadened by an ideal column, but is by a non-ideal column.

column, mixed phase A column containing more than one stationary phase.

column packing The term describing the combined liquid phase and solid support.

column performance An expression which is described in terms of the plate number and relates to the efficiency of the column to separate consecutive components of a sample.

column, polar A column constructed from a stationary phase which preferentially retains solute having polar functional groups.

column, non-polar A column constructed from a stationary phase from which the solutes are eluted according to their boiling points.

component A single compound contained in a mixture.

dead volume The total volume in the gas line within the gas chromatograph which does not contain the stationary phase. The volume comprises the space within the injection block and detector and also the interstitial spaces in the stationary phase.

detector A device which measures the change in the composition of the effluent gases during the elution of the solutes. A *differential* detector measures instantaneous concentrations, whereas an *integral* detector continuously measures the sample accumulated from the beginning of the analysis.

detector, concentration A detector which gives a signal proportional to the amount of solute per unit volume of carrier gas passing through the detector.

detector, mass A detector which gives a signal proportional to the amount of solute passing through the detector in unit time.

detector response factor The proportionality constant which relates the area of the recorded chromatographic peak to the mass of the solute responsible for the peak. For accurate evaluation of percentage composition of a sample, it is convenient to use *relative response factors* based upon the arbitrary assignment of a response factor equal to unity for one of the components of the sample.

dual column operation A technique in which the analyser column is connected in parallel with a reference column. The signals from the reference and analyser detectors have opposite polarity and are fed simultaneously to the recorder. The use of a dual column system is advantageous when the rate of column "bleed" is high.

elution temperature The temperature at which the component emerges from the column during temperature programmed gas chromatography.

gas chromatography All chromatographic methods in which the mobile phase is a gas.

gas chromatography, cryogenic The technique of gas chromatographic separation using a column at sub-ambient temperatures.

gas chromatography, flow programmed A chromatographic technique whereby the elution of the solute from the stationary phase is aided by a gradual increase in the carrier gas flow rate *gas chromatography, preparative*   The gas chromatographic separation of components of a mixture on such a scale as to allow the isolation and collection of the components. The procedure is usually repetitive and involves automatic injection of the sample.

gas chromatography, temperature programmed   A chromatographic technique whereby the elution of the solutes from the stationary phase is aided by a predetermined increase in the column temperature.

gas-liquid chromatography   All chromatographic methods in which separation of the components of a mixture is effected by partition of the sample between a stationary liquid phase and a mobile gas phase.

gas-solid chromatography   All chromatographic methods in which separation of the components of a mixture is effected by partition of the sample between a stationary solid phase and a mobile gas phase.

height equivalent to a theoretical plate   see *plate height*.

Henry's law   The law which describes the distribution of a solute between two immiscible phases.

injector   A device via which the sample is introduced onto the column.

interstitial volume   The volume in a column not occupied by the stationary phase.

isotherm   The relationship, defined by Henry's law, describing the change in concentration of the solute in the liquid phase which accompanies a change in the concentration of the solute in the gas phase.

liquid volume   The volume occupied by the liquid phase in a column.

mobile phase   see *carrier gas*.

partition coefficients   A physico-chemical constant describing the equilibration of the solute between the stationary phase and the carrier gas.

peak   The record of a change in the detector signal during the passage of a component through an integral detector.

pyrolysis chromatography   A procedure in which relatively non-volatile complex compounds may be characterised by the gas chromatogram of their pyrolysis products.

plate height   A term which describes the efficiency of the column by analogy with fractional distillation. The plate height is that length of column which is required to establish perfect equilibration of the solute between the stationary phase and the mobile phase (see also *plate number*).

plate number   The number of theoretical plates required to effect a separation of two consecutive components of the sample.

resolution   A term which describes the separation of consecutive solutes in terms of their band widths and the separation of their maxima.

retention index (Kovats')   A description of the retention of a solute on a column in terms of its adjusted retention volume relative to those of *n*-alkanes. The retention index is expressed as the number of carbon atoms multiplied by 100 of a hypothetical *n*-alkane which under the same conditions would have the same retention volume as that of the solute.

retention, relative   The ratio of the adjusted retention volume or retention time of a solute to that of an arbitrarily chosen reference compound, the adjusted retention volume or retention time of which has been equated to unity.

retention temperature   see *elution temperature*.

retention time   The time required, from the time of its injection, for the elution of the solute from the column.

retention volume   The volume of carrier gas required to carry the solute through the column from its point of injection. The volume may be calculated by multiplying the retention time by the volumetric flow rate of the carrier gas.

retention volume, adjusted   The retention volume of the solute in which a correction has been made for the dead volume of the instrument.

retention volume, net   The adjusted retention volume of a solute in which a correction has been made for the pressure gradient of the carrier gas along the length of the column.

retention volume, specific   The retention volume of a solute which is independent of the amount of liquid phase in the column.

sample   The mixture injected onto the column for chromatographic analysis.

solid support   The inert solid which supports the liquid phase in the column.

solute   Usually refers to that part of the sample which is dissolved or adsorbed in the liquid phase, but may also be used to describe the sample.

solvent flush technique (for sample injection)   A technique whereby accurately measured quantities of sample may be injected into the gas chromatograph.

stationary phase   A relatively nonvolatile liquid absorbed on a solid support where it acts as the liquid phase in gas chromatographic separations.

stream splitter   A device for the division of the carrier gas or of the column effluent into two streams. Incorporated in the injector block for use with capillary columns or, alternatively, fitted between the column and the detector such that the major part of the effluent gases by-pass the detector.

tailing An assymetry in the peak shape in which the gradient of the trailing edge is relatively shallow.

total gas hold-up see *dead volume*.

van Deemter equation An equation which relates the plate height with the many variables associated with the flow of the carrier gas through the stationary phase.

A	eddy diffusion constant in the simplified van Deemter equation
	area of peak
B	longitudinal diffusion coefficient in the simplified van Deemter equation
C	constant in the simplified van Deemter equation dependent upon the non-instantaneous equilibration of the solute between the gas and liquid phases
D_g	diffusion coefficient of the solute in the gas phase
D_l	diffusion coefficient of the solute in the liquid phase
F_n, F_a, F_d	retention fractions calculated from retention volume data from non-polar, electron accepting and electron donating polar columns
H	the height of the triangle enclosing a peak and constructed by the intercepts of the baseline and the tangent lines.
ΔH_v	enthalpy of vaporisation of the solute from the stationary phase
I	retention index
K	peak area correction factor for flame ionisation detectors
M_c	molecular weight of the solute
M_g	molecular weight of the carrier gas
P_i	pressure of carrier gas at column inlet
P_o	pressure of carrier gas at column outlet
R	column resolution
T_c	column temperature
T_f	flowmeter temperature
T_I	initial column temperature prior to temperature programmed gas chromatography
T_R	elution or retention temperature
V_n, V_a, V_d	adjusted retention volumes of a solute measured on non-polar, electron accepting and electron donating columns
V_g	specific retention volume
V_M	dead volume (total gas hold-up of the column)
V_N	net retention volume
V_R	measured retention volume
$V_{R'}$	adjusted retention volume

$V_R{}^0$	corrected retention volume
W	total weight of stationary phase in the column
	base width of peak estimated by triangulation technique
$W_{h/2}$	peak width at half height
a	volume of gas per unit length of column
c	concentration of the solute in the gas phase
f	detector response factor
h	height equivalent to a theoretical plate
	measured peak height
j	pressure correction factor
l	length of column
m	mass of stationary phase per unit length of column
	mass of component passing through the detector
q	concentration of solute in the stationary phase
r	relative retention
	column radius
t_R	retention time
u	programmed heating rate of column during temperature programmed gas chromatography
v	flow rate of the carrier gas
w	base width of peak
α	dimensionless partition coefficient
β	partition coefficient with the dimension ml^{-3}
γ	partition coefficient with the dimension ml^{-3}
ρ	density of the stationary phase
σ	standard deviation of a peak
BSA	N,O-bistrimethylsilylacetamide
BSFA	N,O-bistrimethylsilyltrifluoroacetamide
BSS	British standard sieve
CAT	computer average transients
EC	electron capture (detector)
FID	flame ionisation detector
GLC	gas-liquid chromatography
HETP	height equivalent to a theoretical plate
HMDS	hexamethyldisilazane
IR	infrared spectroscopy
MS	mass spectrometry
NMR	nuclear magnetic resonance spectrometry
TC	thermal conductivity (detector)
TLC	thin layer chromatography
TMCS	trimethylchlorosilane
TMDA	trimethylsilyldiethylamine

B. Operational Procedures and Common Operational Faults

OPERATION OF A GAS CHROMATOGRAPH FITTED WITH FLAME IONISATION AND/OR THERMAL CONDUCTIVITY DETECTORS

The following list of operations is intended as a general guide for the preparation of a typical gas chromatograph for simple routine analysis. The procedure may vary slightly depending upon the design of the instrument and the operator is advised to consult the instrument manual for precise information.

1. Switch on mains supply to instrument.
2. Switch on detector amplifier (FID only). With some instruments this switch also controls the current to the detector electrodes (see 7(a) below).
3. Switch on recorder unit.
4. Fit selected column. (With FID select effluent stream splitter ratio.)
5. Set external carrier gas supply pressure to at least 10 lb/in² above the proposed operational pressure and adjust flow rate to desired value (c. 25–50 ml/min for ¼ in standard packed column).
6. Check instrument for gas leaks.
7. Switch on heater power to TC detector and allow system to reach equilibrium.
   OR
   (a) Switch on power supply to FID.
   (b) Adjust H_2 and air supplies to FID and ignite flame.
8. Balance FID amplifier (electrometer). [It is usually preferable with some instruments to balance the amplifier before flame ignition.]
9. Set oven temperature control to required operational temperature.
10. Select required injection block temperature.
11. Check oven temperature and ensure that thermal equilibrium has been attained.
12. Check carrier gas flow rate and adjust if necessary.
13. Adjust filament current (TC) for maximum sensitivity.

14. Set attenuator to zero and adjust base line on recorder to zero. [To make these adjustments when the gas chromatograph is fitted with a TC detector it may be necessary to adjust the Bridge balance controls.]
15. Inject sample.

COMMON OPERATIONAL FAULTS

The following notes will help the inexperienced chromatographer to detect and correct simple operational faults and to obtain a better performance from the instrument. These notes should, of course, be used in conjunction with the instrument manual. The first section describes the most common causes of poor response and the second section summarizes the general causes of poor peak profiles.

INSTRUMENT RESPONSE

Symptom	Possible cause	Remedy
No peaks	No carrier gas flow	Check that the carrier gas flow is on and adjusted to the correct flow rate. Replace gas cylinder if it is empty. Check that all column connections are tight. Check all gas lines for obstructions.
	Detector not operating	Check that the detector power switch is on and that the detector is adjusted to a suitable sensitivity level. (FID only) Check that the flame is alight. If necessary check air and hydrogen gas pressures. (Ionisation detectors) Check that the cell voltage is on. Check all electrical connections.
	Defective recorder	Consult recorder instruction manual.
	Sample not being vaporised in the injection block	Increase injection block temperature.
	Sample condensing on column	Increase column temperature.
	Ineffective sample introduction	Check that syringe is operating adequately. Check whether the injector septum is leaking. Replace if necessary.

COMMON OPERATIONAL FAULTS—*Continued*

Symptom	Possible cause	Remedy
Poor sensitivity (low peak heights)	Incorrect attenuation	Reduce attenuation. Set Range Switch for ionisation detectors to a lower value.
	Insufficient sample reaching the detector	Increase sample size. Check for ineffective sample introduction. Check for carrier gas leak.
	Low detector response	(TC only) Increase filament current or operate at lower temperature. (FID only) Increase flame size by increasing hydrogen and/or air pressures. Clean the electrodes.
	Peak broadening due to extended retention time	Increase carrier gas flow and/or column temperature.
Baseline cannot be zeroed	Recorder incorrectly adjusted or connected	Consult recorder instrument manual.
	Defective detector	(TC only) Check power supply. Check that the filaments are balanced. Replace if necessary. (Ionisation detectors) Clean electrodes.
	High signal from excessive column "bleed"	Reduce column temperature. Change column.
High noise level	Poor electrical connections	Check that all earth connections are good. Check that all plugs are properly seated and that the connections are good. Check for dirty switch contacts. Clean if necessary with a contact cleaner.

B. OPERATIONAL PROCEDURES AND COMMON OPERATIONAL FAULTS

F	Dirty recorder slide wire	Noise usually appears in the same area on the chart. Rectify by cleaning the recorder slide wire with CCl_4. The recorder may be defective. This may be checked by shorting the recorder input. If the noise continues, consult the recorder manual.
	Contaminated carrier gas	Replace carrier gas filter. Check for leaks in the gas line. Noise can also result when the carrier gas flow rate is too high.
	Contaminated column	Change or recondition column. Excessive column "bleeding" can also produce noise.
	Dirty detector	(TC) Clean detector block. If detector filaments highly contaminated, change detector. (Ionisation detectors) Clean insulators. Water condensing on the inside of flame ionization detectors can result in a high noise level and may be obviated by increasing detector temperature above 100°.
	Contaminated hydrogen and air for flame ionisation detector.	Replace filters. Incorrectly adjusted flow rates can also lead to noise.
	Contaminated injector assembly.	Dismantle and clean. Replace septum.
Regular or irregular sharp peaks ("spiking")	Mains voltage fluctuations	Use an isolated electrical output and/or a stabilised transformer. Defective power supply to TC detectors can also produce spiked peaks.
	Rapid change in atmospheric pressure and/or temperature	Locate instrument away from draughts, heaters, etc.
	Intermittant shorting in ionisation detectors	Clean insulators and connectors.
		(FID only) Dust within the detector may find its way into the flame. Clean the detector by blowing out with a stream of clean air or nitrogen.

COMMON OPERATIONAL FAULTS—*Continued*

Symptom	Possible cause	Remedy
Drifting base line (a) Steady baseline drift in one direction	Detector temperature not constant	Allow time for detector to come to a stable temperature (This is particularly important for TC detectors). Check temperature control.
	Defective detector circuit	Consult instrument manual. Thermal conductivity detectors are susceptible to air leaks. If the leak cannot be found, the defect may be obviated by increasing the carrier gas flow rate. Check that both filaments and their power supply circuits are fully efficient (see manual). Change if necessary.
	Increase in column "bleeding" with increasing temperature (Temperature programming only)	Use dual column system (see p. 119). Use lower ratio of liquid phase on column support (see p. 64).
	Carrier gas flow rates not balanced (Dual column system only)	Adjust flow rates as directed in instrument manual.
(b) Irregular baseline drift	Instrument is being subjected to excessive changes in the ambient temperature.	Relocate instrument away from draughts, heaters, etc.
	Excessive column "bleeding".	Operate at a lower temperature (see p. 66). Use a column with a lower ratio of stationary phase to support (see p. 64). Use dual column system (see p. 119). However if the columns are not properly conditioned an irregular baseline will still be produced (see p. 34 and consult manual for the recommended procedure for conditioning column).

B. OPERATIONAL PROCEDURES AND COMMON OPERATIONAL FAULTS 149

Irregular carrier gas flow	Adjust carrier gas flow rate for optimum performance. For dual cloumn operation, it is important that the flow rates are balanced (see p. 120).
	Check all gas lines for leaks.
	Check carrier gas regulator. When the gas cylinder is almost empty there is insufficient gas pressure for the regulator to operate efficiently.
Dirty or defective detector	Clean or replace detector (see above). Ensure that detector oven temperature is stable.
Defective detector circuit	Consult instrument manual.
(FID only) Irregular flame	Check hydrogen and air flow to ensure optimum conditions.
Defective recorder	See above and consult recorder manual. Generally check both recorder and chromatograph to ensure good electrical connections.
(c) Base line "stepping"	
Recorder gain and/or damping control incorrectly adjusted	With reference to recorder manual adjust controls. One should not be able to move the recorder pen easily with finger pressure when these controls are correctly adjusted.
Dirty recorder slide wire	Clean with CCl_4.
Recorder or chromatograph not connected to a good earth.	Check and rectify if necessary.
(d) Sinusoidal baseline	
Column oven control defective	Check and replace if necessary.
Detector temperature control defective	Check and replace if necessary.
Carrier gas flow regulator defective	Check and replace if necessary. The regulator will not operate efficiently when the gas cylinder is almost empty.

COMMON OPERATIONAL FAULTS—Continued

Symptom	Possible cause	Remedy
Negative peaks	Input leads to recorder reversed	Reconnect leads correctly.
	Detector polarity switch in wrong position	Switch to opposite polarity.
	Sample injected onto wrong column (Dual column operation)	Flush out column and inject sample into analytical column.
	(Electron Capture detector) If negative signals appear after peaks, the detector is contaminated	Clean detector according to instruction manual.
	Sample size too large (loss of detector sensitivity)	Reduce sample size.
Negative signal on recorder and FID flame extinguished	Carrier gas flow rate too high	Readjust flow rate.
	Hydrogen or air flow impeded	Check and readjust.
	Blocked flame jet	Clean or replace.
	Sample size too large	Reduce sample size.

Peak Profile

Asymmetric peaks

(a) Tailing peaks (see Fig. 2b, p. 6)	Column temperature too low	Increase column temperature.
	Column overloaded under conditions where sample is completely vaporised	Reduce sample size (see p. 10).

B. OPERATIONAL PROCEDURES AND COMMON OPERATIONAL FAULTS 151

	Sample injection is too slow	This is usually only important when the injection block temperature is too low to permit flash evaporation of the sample (see p. 23).
	Injection block temperature too low	Adjust temperature to optimum performance (see p. 23).
	Sample interacting with column solid support	Condition column and/or treat column with a silylating agent (see p. 32).
	Sample interacting strongly with column liquid phase	Change column.
(b) Leading peaks (see Fig. 2c, p. 6)	Column overloaded under conditions where the sample is not completely vaporised	Increase column temperature and/or reduce sample size.
	Sample condensed in injection port and/or column	Adjust temperatures to optimum performance (see p. 23).
Unresolved peaks	Column overloaded by major components of mixture (see p. 10)	Reduce sample size.
	Column temperature too high and/or carrier gas flow rate too high	Adjust temperature and/or flow rates to give optimum separation of peaks.
	Column too short	Use longer column.
	Incorrect choice of column liquid phase (see p. 159)	Change column.
	Sample injection is too slow resulting in broadened input distribution of sample	Improve injection technique.

COMMON OPERATIONAL FAULTS—*Continued*

Symptom	Possible cause	Remedy
Round topped peaks	Recorder gain too low	Consult recorder manual and adjust gain settings.
	Sample size outside the dynamic range of detector output (this can also produce flat topped peaks)	Reduce sample size.
Flat topped peaks	Recorder gain and/or damping control incorrectly adjusted (this is usually accompanied with stepping)	Consult recorder manual and adjust controls.
	100% Deflection on recorder incorrectly adjusted	Consult recorder manual and adjust controls.

C. Stationary Phases in General Use

Virtually any inert organic liquid, or low m.p. solid, of low volatility is potentially suitable as a liquid stationary phase for gas-liquid chromatography. The commonly used stationary phases are listed alphabetically in Table VI and they are classified in Table VII according to their application in the separation of specific classses of organic compounds.

TABLE VI

Stationary phase	Polarity*	Maximum operational temperature	Uses
Apiezon (types H to N)	NP	Variable (up to 300)	General purpose high temperature separations.
Alkyl aryl sulphonates (e.g. Tide)	SP	c. 250	General purpose semi-polar phase ready for pouring directly into column tube.
p,p'-Azoxyphenetole			Liquid crystalline compound used for the separation of isomeric arenes.
Bentone 34	see dinonyl phthalate		
Benzyl cellosolve	SP		SO_2, H_2O, H_2S, CO_2.
Benzyl cyanide- silver nitrate		50	Specific for the separation of alkenes.
Benzyl diphenyl	NP	120	Arenes.

TABLE VI—*Continued*

Stationary phase	Polarity*	Maximum operational temperature	Uses
Bis-phenetidylterphthalaldehyde			Liquid crystalline compound used for the separation of isomeric arenes.
Carbowax 400		100	General purpose polar phase.
550			
750	Polarity decreases with M.W.		
1000			
1500			
1540			
4000			
6000			
20M		→ 250	
Cholesteryl benzoate			Liquid crystalline compound used for the separation of isomeric arenes.
Cyanoethylsilicones (e.g. XE-60, XF-1150, ECNSS-S, ECNSS-M)	P	c. 200	Ketones, esters, amines, alcohols, steroids.
CYANO-B	P	200	Diastereoisomeric esters.
Dibenzyl ether	SP	80	Chloro compounds.
Dibutyl phthalate	SP	100	Fluoro compounds.
Didecyl phthalate	SP	175	Hydrocarbons.
Di(2-ethylhexyl) sebacate	SP	150	General purpose semi-polar phase.
Di(2-ethylhexyl) phthalate	SP	150	Fluoro compounds, aldehydes, alkanes, arenes, mercaptans.
Diglycerol	P	150	Alcohols, aldehydes.

C. STATIONARY PHASES IN GENERAL USE 155

Di-isodecyl phthalate	SP	175	General purpose semi-polar phase.
Di(2-methoxyethyl) adipate	P	100	Low M.W. alkanes.
Dimethyl formamide	P	50	Low M.W. alkanes.
Dimethyl stearamide	SP	150	Alcohols, aldehydes, ketones.
Dimethyl sulpholane	P	50	Alkenes.
Dimethyl sulphoxide	P	50	CO_2, N_2O, O_2.
Dinonyl phthalate	SP	150	Ethers, esters, ketones, arenes, sulphur compounds.
Dioctyl phthalate	SP	160	General purpose semi-polar phase.
Di(3,3,5-trimethylhexyl)phthalate	SP	140	Phenols.
EGSS-X and EGSS-Y			Copolymer of ethylene succinate and methyl siloxane.
EGSP-A and EGSP-Z			Copolymer of ethylene succinate and phenyl siloxane.
ECNSS-S and ECNSS-M			Copolymer of ethylene succinate and cyanoethyl siloxane.
Ethylene glycol—silver nitrate		50	Specific for alkenes.
Fluorosilicones (e.g. FS-96, FS-1265, QF-1)	P	250	Alcohols, halogeno compounds, steroids.
Glycerol	P	100	Alcohols.
Hallcomid M-18			See dimethyl stearamide
Hexamethylphosphoramide	P	35	Low M.W. alkenes.
Igepal 880			See nonylphenoxypoly(ethyleneoxy)ethanol.
LAC 1-R-296			See polypropylene glycol adipate.
LAC 2-R-446			See polydiethylene glycol adipate.
LAC 3-R-728			See polydiethylene glycol succinate.
LAC 6-R-860			See poly-1,4-butanediol succinate.
Nonylphenoxypoly(ethyleneoxy)-ethanol	P	200	Chloroarenes.

TABLE VI—*Continued*

Stationary phase	polarity*	Maximum operational temperature	Uses
Octylphenoxypoly(ethyleneoxy)-ethanol	P	200	Sulphur compounds.
β,β'-Oxydipropionitrile	SP	100	Alkenes.
Perfluorocarbons (Freons)	P	Variable	Useful for the separation of hydrocarbons from fluoro-compounds.
Phenyl diethanolamine succinate	P	225	Aldehydes, ketones, esters, nitriles.
Polyamide resins	See Versamid		
Poly-1,4-butanediol succinate	SP	200	Esters.
Polyethylene glycol	See Carbowax		
Polyethylene glycol—thallium nitrate		100	Specific for alkenes.
Polyethylene glycol adipate	SP	225	Terpenes, alcohols.
Polyethylene glycol isophthalate	SP	190	Esters.
Polyethylene glycol succinate	SP	225	Esters.
Polydiethylene glycol adipate	SP	225	Esters, terpenes.
Polydiethylene glycol succinate	SP	225	Esters, ethers, terpenes.
Polyphenyl ethers (e.g. OS-124 and OS-138)	SP	200	Polyarenes, haloarenes.
Polypropylene glycols	P	Variable up to 150	General purpose polar phase.
Polypropylene glycol adipate	SP	200	Esters.

C. STATIONARY PHASES IN GENERAL USE 157

Propylene carbonate		SP	50	CO_2, N_2O, O_2.
Quadrol		P	150	Amines, aldehydes.
Silicone oils and greases		NP		General purpose, non-polar phase.
DC-200	methyl siloxane		250	
DC-410	methyl siloxane		300	
DC-430	dimethylvinyl siloxane		300	
DC-550	methyl phenyl siloxane		225	
DC-710	methyl phenyl siloxane		225	
GE SE-52	methyl phenyl siloxane		300	
GE SE-54	phenyl vinyl siloxane		300	
GE SE-96	methyl siloxane		250	
GE SE-30	methyl siloxane		300	
GE SE-33	methyl siloxane		300	
OV-1	methyl siloxane		350	
OV-25	methyl phenyl siloxane		300	
OV-17	methyl phenyl siloxane		350	
UC L-45	methyl siloxane		300	
UC W-98	methyl vinyl siloxane		300	
Squalane		NP	150	Alkanes.
Tetraethylenepentamine		P	60	Ammonia and gaseous amines.
1,2,3,4-Tetrakis(2-cyanoethoxy)-butane		See CYANO-B.		
N,N,N',N'-Tetrakis[2(2'-cyanoethoxy)propyl]ethyleneamine		P		Halogeno compounds.
N,N,N',N'-Tetrakis(2-hydroxypropyl)ethylenediamine		See Quadrol		
Tricresyl phosphate		SP	125	Alkanes, alcohols, ethers, sulphur compounds.

158 INTRODUCTION TO GAS-LIQUID CHROMATOGRAPHY

TABLE VI—*Continued*

Stationary phase	Polarity*	Maximum operational temperature	Uses
Tricyanopropane	P		Used for the separation of (+) and (−) camphor via its diastereoisomeric 2,3-butane diols.
Tritolyl phosphate	See tricresyl phosphate		
Triton X-100 and X-305	See octylphenoxypoly(ethyleneoxy)ethanol		
UCON fluids (prefix LB—water insoluble prefix HB or H—water soluble)	See polypropylene glycols		
Versamid 930	SP	250	General purpose for high boiling polar compounds.

* NP non-polar; SP semi-polar; P polar.

C. STATIONARY PHASES IN GENERAL USE

TABLE VII

Stationary phase	Maximum operational temperature	
General purpose		
High molecular weight hydrocarbons (Apiezon)	< 300	Non-polar, useful for high temperature work.
Silicone oils	< 200	Non-polar
Silicone greases	300–350	Non-polar
Di-(2-ethylhexyl) sebacate	150	Semi-polar
Di-isodecyl phthalate	150	Semi-polar
Polyethylene glycols (Carbowaxes) (M.W. 200–20,000)	< 250	Polarity is inversely proportional to M.W.
Polypropylene glycols (Ucon fluids)	< 150	Polar
Separation of Hydrocarbons		
Di(2-methoxyethyl) adipate + di-(2-ethylhexyl) sebacate (2 : 1)	80	Suitable for gaseous alkanes, polar.
Di(2-methoxyethyl) adipate	100	Suitable for low M.W. alkanes, polar.
Squalane	150	Suitable for alkanes and arenes.
Silicone oils	< 200	Suitable for alkanes.
Dialkyl phthalates	< 175	Suitable for alkanes, alkenes and arenes.
Dimethylformamide	50	Suitable only for low M.W. alkanes.
Tricresyl phosphate	125	Suitable for alkanes and arenes, toxic.
Hexamethylphosphoramide	35	Suitable for low M.W. alkenes, polar.
Dimethylsulpholane	50	Suitable for alkenes, polar.
β,β'-Oxydipropionitrile	100	Suitable for alkenes, polar.
Silver nitrate—ethylene glycol	50	Specific for alkenes.
Silver nitrate—benzyl cyanide	50	Specific for alkenes.
Thallium nitrate—polyethylene glycol	100	Specific for alkenes.
Benzyl diphenyl	120	Suitable for arenes and alkanes.
Polypropylene glycols (Ucon fluids)	< 150	Suitable for arenes.
Polyethylene glycols (Carbowaxes)	< 250	Suitable for arenes.
Separation of Ketones and Aldehydes		
Polyethylene glycols (Carbowaxes)	100–250	Polar
Diglycerol	150	Hydrogen bonding polar phase.
N,N,N',N'-Tetrakis(2-hydroxypropyl) ethylenediamine (Quadrol)	150	Polar

TABLE VII—*Continued*

Stationary phase	Maximum operational temperature	
Dialkyl phthalates	< 175	Semi-polar
Cyanosilicones (XF-1150)	200	Polar
Polypropylene glycol (Ucon LB-500-X)	200	Polar
Phenyldiethanolamine succinate	225	Polar
Fluorosilicone oils (FS-96)	250	Polar, suitable for high b.p. carbonyl compounds.

Separation of Esters

Tricresyl phosphate	125	Recommended for acetates, toxic.
Dialkyl phthalates	< 175	Semi-polar
Cyanosilicones (XF-1150)	200	Polar
Silicone oils, greases and gums	200–350	Non-polar
Polydiethyleneglycol succinate (LAC 3-R-728)	200	Polar, recommended for methyl esters.
Polydiethyleneglycol adipate (LAC 2-R-446)	200	Polar
Poly-1,4-butanediol succinate (LAC 6-R-860)	200	Polar
Polyethylene glycol (Carbowax 20M)	250	Polar, suitable for high b.p. esters.
High molecular weight hydrocarbons (Apiezon)	< 300	Non-polar, suitable for high b.p. esters.

Separation of Alcohols and Phenols

Silicone oils and greases	< 300	Non-polar suitable for alcohols.
Polyethylene glycol (Carbowax 400)	100	Polar, suitable for alcohols.
Polypropylene glycol (Ucon LB-500-X)	200	Polar, suitable for alcohols.
Fluorosilicone oils (FS-1265)	225	Polar, suitable for high b.p. alcohols.
Glycerol	100	Hydrogen bonding phase, suitable for alcohols.
Diglycerol	150	Less polar than glycerol, suitable for alcohols.
Di(3,3,5-trimethylhexyl) phthalate	140	Semi-polar, suitable for phenols.
Cyanosilicones (XF-1150)	200	Polar, suitable for phenols and alcohols.
Phenyldiethanolamine succinate	225	Polar, suitable for phenols.

C. STATIONARY PHASES IN GENERAL USE

TABLE VII—*Continued*

Stationary phase	Maximum operational temperature	
Polyethylene glycol (Carbowax 20M)	250	Polar, suitable for phenols and high b.p. alcohols.
Silicone gums	400	Non-polar, suitable for high b.p. phenols.
Separation of Ethers		
Polyethylene glycols (Carbowax 400)	100	Polar
Tricresyl phosphate	125	Semi-polar, toxic.
Dialkyl phthalates	< 175	Semi-polar
Polydiethyleneglycol succinate (LAC 3-R-728)	200	Polar
High molecular weight hydrocarbons (Apiezon)	< 300	Non-polar
Separation of Amines		
Cyanosilicones (XF-1150)	200	Polar
Polyethylene glycol on KOH treated support	250	Basic and polar.
Apiezon on KOH treated support	< 300	Basic and non-polar.
Polyamide resins (Versamid) on KOH treated support	< 350	Semi-polar, suitable for high b.p. amines.
Separation of Halogeno compounds		
Dibenzyl ether	80	Suitable for chloro compounds.
Dialkyl phthalates	< 175	Semi-polar, suitable for chloro compounds.
Silicone oils	< 200	Non-polar, general use.
Fluorosilicone oils (QF-1)	250	Polar, general use.
Perfluorocarbons (Freons)		Retain fluoro compounds have no retention for alkanes.
Polyethylene glycols (Carbowaxes)	< 250	Polar, general use.
High molecular weight hydrocarbons (Apiezon)	< 300	Non-polar, general use for high b.p. compounds.
Nonylphenoxypoly(ethyleneoxy)-ethanol (Igepal 880)	200	Recommended for aromatic chloro compounds.
Separation of Sulphur compounds		
Tricresyl phosphate	125	Semi-polar, suitable for mercaptans.
Squalane	150	Non-polar, suitable for sulphides.

TABLE VII—*Continued*

Stationary phase	Maximum operational temperature	
Dialkyl phthalates	< 175	Semi-polar, suitable for mercaptans and sulphides.
Silicone oils	< 200	Non-polar, suitable for sulphides.
Separation of Essential Oils		
N,N,N′,N′-Tetrakis(2-hydroxypropyl)ethylenediamine (Quadrol)	150	Polar
Cyanosilicones (XF-1150)	200	Polar
Polydiethyleneglycol succinate (LAC 3-R-728)	200	Polar
Polydiethyleneglycol adipate (LAC 6-R-446)	200	Polar
Polypropylene glycols (Ucon fluids)	< 225	Polar
Phenyldiethanolamine succinate	250	Polar
High molecular weight hydrocarbons (Apiezon)	< 300	Non-polar
Separation of Steroids and Alkaloids		
Cyanosilicones (XE-60)	200	Polar, particularly suitable for steroids.
Fluorosilicones (QF-1)	250	Polar
Silicone gums	< 400	Non-polar
Separation of Gases		
Dimethyl sulphoxide	50	Suitable for CO_2, N_2O, and O_2.
Propylene carbonate	50	Suitable for CO_2, N_2O, and O_2.
Benzyl cellosolve		Suitable for SO_2, H_2O, H_2S, and CO_2.
Tetraethylenepentamine	60	Suitable for NH_3 and amines.
N,N,N′,N′-Tetrakis(2-hydroxypropyl)ethylenediamine (Quadrol) on KOH treated support	150	Suitable for NH_3 and amines.
Di(2-methoxyethyl) adipate + di-(2-ethylhexyl) sebacate (2 : 1)	80	Suitable for gaseous hydrocarbons.
Separation of Geometric Isomers		
Silver nitrate—ethylene glycol	50	Specific for *cis*- and *trans*-alkenes.
Silver nitrate—benzyl cyanide	50	Specific for *cis*- and *trans*-alkenes.

C. STATIONARY PHASES IN GENERAL USE

TABLE VII—*Continued*

Stationary phase	Maximum operational temperature	
Thallium nitrate—polyethylene glycol	100	Specific for *cis*- and *trans*-alkenes.

Liquid crystals[1]

p,p'-Azoxyphenetole (nematic temperature range 138–168°)		Used for the separation of alkylarenes.
Bis-phenetidylterphthalaldehyde		Used for the separation of arenedicarboxylic esters, dimethoxybenzenes, α- and β-naphthols.
Cholesteryl benzoate		Used for the separation of chlorotoluenes.

Separation of Enantiomers and Diastereoisomers

Silicone oil		Used for the separation of diastereoisomeric alkenes[2] and of N-chloroalkanoylaminoacid esters[3].
Squalane		Used for the separation of diastereoisomeric alkanes[4] and of diastereoisomeric α-alkanoyloxypropionic esters[5].
Diglycerol : polyethylene glycol (9 : 1) *or* polypropylene glycol *or* polyethylene glycol adipate *or* tricresyl phosphate		Used for the separation of diastereoisomeric alkanols[6,7].
Tricyanopropane		Used for the separation of (+) and (−) camphor via the diastereoisomeric 2,3-butanediol ketals[8].
1,2,3-Tri(2-cyanoethoxy)-propane		Used for the separation of diastereoisomeric esters of sec. alkanols[9].
1,2,3,4-Tetra(2-cyanoethoxy)-butane		Used for the separation of diastereoisomeric esters of 2,3-butanediol[10].
N,N,N',N'-Tetrakis[2(2'-cyanoethoxy)propyl]ethylene amine		Used for the separation of diastereoisomeric halogeno compounds[11].
Alkyl aryl sulphonates (Tide)		Used for 'the separation of diastereoisomeric bromoalkanes[12].

TABLE VII—*Continued*

Stationary phase	Maximum operational temperature	
Silicone gum (SE-30): polyethylene glycol isophthalate		Used for the separation of diastereoisomeric esters[13].
Polypropylene glycol (Ucon LB 550-X)		Used for the separation of the diastereoisomers of alkanoyloxy esters[5, 14] and of N-trifluoroacetyl derivatives of aminoacid esters and dipeptide esters[5, 15, 16, 17].
Polyethylene glycol (Carbowaxes 1540 and 20M)		Used for the separation of diastereoisomeric N-trifluoroacetyl derivatives of aminoacid esters of 2-n-alkanols[15, 16].
Polydiethyleneglycol succinate : methylsiloxane (EGSS-X)		Used for the separation of diastereoisomeric N-trifluoracetyl-L-prolyl derivatives of D and L amines[18] and of N-trifluoroacetyl derivatives of diastereoisomeric aminoacid esters[19].
Polyethyleneglycol succinate		Used for the separation of N-α-chloroisovaleryl derivatives of D and L aminoacids[20].
Poly-1,4-butanediol succinate		Used for the separation of diastereoisomeric N-trifluoracetyl aminoacid esters of 2-n-alkanols[5].
Fluorosiloxanes (FS-1265)		Used for the separation of diastereoisomeric derivatives of aminoacid esters and α-alkanoyl oxypropionates[5].
Polyphenyl ethers (OS-138)		Used for the separation of diastereoisomeric N-trifluoracetyl derivatives of dipeptides[17].
Ureide of L valine isopropyl ester		Used for the separation of enantiomeric N-trifluoacetyl derivatives of primary amines[21].
N-Trifluoacetyl-L-isoleucine lauryl ester *or* N-trifluoracetyl-L-phenyl alanine cyclohexyl ester		Used for the separation of enantiomeric aminoacid esters[22, 23].

REFERENCES FOR TABLE VII

1. The three examples given illustrate the typical structures of compounds which have an ordered liquid phase. For a detailed review describing the uses of liquid crystals as stationary phases for gas chromatographic separation see Kelker, H. and von Schivizhoffen, E. *In* "Advances in Chromatography" (J. C. Giddings and R. A. Keller, eds.) Vol. 6, p. 247, Arnold, London (1968).
2. Doering, W. E. and Roth, W. R., *Tetrahedron*, **18**, 67 (1962).
3. Halpern, B., Westley, J. W. and Weinstein, B., *Nature*, **210**, 837 (1966).
4. Simmons, M. C., Richardson, D. B. and Dvoretzky, I. *In* "Gas Chromatography, 1960" (R. P. W. Scott, ed.) p. 211, Butterworths, London (1960).
5. Gil-Av, E., Charles-Sigler, R., Fischer, G. and Nurok, D., *J. Gas. Chromatog.*, **4**, 51 (1966).
6. Arens, C. L., Cort, L. A., Howard, T. J. and Loc, L. B., *J. Chem. Soc.*, p. 1195 (1960).
7. Gault, Y. and Felkin, H., *Bull. Soc. chim. France*, p. 742 (1965).
8. Cassanova, J. and Corey, E. J., *Chem. and Ind.*, p. 1664 (1961).
9. Rose, H. C., Stern, R. L. and Karger, B. L., *Anal. Chem.*, **38**, 469 (1966).
10. Nurok, D., Taylor, G. L. and Stephen, A. M., *J. Chem. Soc. (Section B)*, p. 291 (1968).
11. Stern, R., Atkinson, E. R. and Jennings, F. C., *Chem. and Ind.*, p. 1758 (1962).
12. Goering, H. L. and Larson, D. W., *J. Am. Chem. Soc.*, **81**, 5937 (1959).
13. Stern, R. L., Karger, B. L., Keane, W. J. and Rose, H. C., *J. Chromatog.*, **39**, 17 (1969).
14. Gil-Av, E. and Nurok, D., *Proc. Chem. Soc.*, p. 146 (1962).
15. Pollock, G. E. and Oyama, V. I., *J. Gas. Chromatog.*, **4**, 126 (1966).
16. Pollock, G. E., Oyama, V. I. and Johnson, R. D., *J. Gas Chromatog.*, **3**, 174 (1965).
17. Weygand, F., Prox, A., Schmidhammer, L. and König, W., *Angew. Chem.*, **73**, 282 (1963) [*Internat. Edn.* **3**, 183 (1963)].
18. Halpern, B. and Westley, J. W., *Chem. Comm.*, p. 34 (1966).
19. Westley, J. W., Halpern, B. and Karger, B. L., *Anal. Chem.*, **40**, 2046 (1968).
20. Halpern, B. and Westley, J. W., *Chem. Comm.*, p. 246 (1965).
21. Feibush, B. and Gil-Av, E., *J. Gas. Chromatog.*, **5**, 257 (1967).
22. Gil-Av, E., Feibush, B. and Charles-Sigler, R., *Tetrahedron Letters*, p. 1009 (1966).
23. Gil-Av, E., Feibush, B. and Charles-Sigler, R. *In* "Gas Chromatography, 1966" (A. B. Littlewood, ed.) p. 227, Institute of Petroleum, London (1967).

D. Instrument

The selection of the most suitable chromatograph for use in an organic chemistry
In order to help a potential user, the following two tables summarise the construction, selection of currently produced analytical and preparative chromatographs. Further pp. 188 to 191.

The company directory, which follows the instrument directory, indicates the range
A more comprehensive compilation of manufacturers and suppliers of gas chromato-

Key to abbreviations

TC—Thermal conductivity detector; FID—Flame ionisation detector;
Ar—Argon ionisation detector; MCS—Micro cross section detector;
US—Ultrasonic detector; GD—Gas density balance.

RESEARCH AND ROUTINE ANALYTICAL

Manufacturer and model	General description	Instrument dimensions (h × d × l)	Oven dimensions
Barber-Colman Co.			
Series 5000	Dual column research analytical chromatograph. Seven models with choice of detector.	Dimensions vary with model. Consult manufacturer's literature.	1300 cu in
Series 5300	Routine quality control chromatograph for clinical and industrial use.		
Beckman Instruments			
Series GC-M	Model 1240: Dual column research analytical chromatograph.	$19\frac{3}{4} \times 16\frac{1}{8} \times 32\frac{3}{4}$ in (50 × 41 × 83 cm)	10 × 10 × 5 in (25 × 25 × 12 cm)

Directory

laboratory depends to a large extent upon the varied requirements of the operators.
versatility, and the general specifications, as given by the manufacturer, for a wide
information can be obtained from the manufacturers whose addresses are listed on

of instruments and accessories available from the major manufacturing companies.
graphic equipment is to be found in the American Chemical Society Laboratory Guide.

EC—Electron capture detector; Therm.—Thermionic detector;
He—Helium ionisation detector; Photo—Flame photometric detector;

GAS CHROMATOGRAPHS

Detectors	Specifications of basic instruments	Accessories
Choice of TC, FID, EC, Therm.	Isothermal oven operation over range from ambient to 500° with possibility of subambient operation to −180°. Carrier gas flow control over range 5 to 500 ml per min with an accuracy of ±0·3%.	Cryogenic unit operable to −180°. Pyrolyser unit operable over range 300 to 1400°. Gas sampling valve. Automatic time programmed attenuator. Collection unit for preparative chromatography. Cool down controller for re-equilibration of column temperature to initial conditions. Radioactivity monitor for ^{14}C and ^{3}H. Automatic injection system.
Choice of TC, FID, EC, Ar, Therm.	Isothermal and manual or automatic (optional) temperature programmed oven control. On-column injection onto glass columns.	
FID, TC.	Maximum oven temperature of 500° with isothermal and linear temperature programmed operation. Eight heating rates from 1° to 25° per min. Oven stability better than ±0·15°. Flow regulator controls carrier gas flow rates during programmed operation.	Pressure regulators for carrier and fuel gases and flow controller for carrier gas for dual column operation with temperature programming. Micro-thermal conductivity detector and electron capture detector. Three-way valve for dual column operation permitting consecutive flow of sample through two columns or through either one of two columns. Temperature programmer for isothermal models. On-column injection system with ancillary temperature control of inlet. Gas sampling valves.

168   INTRODUCTION TO GAS-LIQUID CHROMATOGRAPHY

RESEARCH AND ROUTINE ANALYTICAL

Manufacturer and model	General description	Instrument dimensions (h × d × l)	Oven dimensions
	Model 1241: Dual column routine chromatograph.	$19\frac{3}{4} \times 16\frac{1}{8} \times 32\frac{3}{4}$ in (50 × 41 × 83 cm)	10 × 10 × 5 in (25 × 25 × 12 cm)
	Steroid analyser.	$19\frac{1}{4} \times 16\frac{1}{8} \times 19\frac{5}{16}$ in (50 × 41 × 49 cm)	10 × 10 × 5 in (25 × 25 × 12 cm)
Series GC-5	High performance analytical dual column instrument. Modular design. Permits 26 standard configurations.	$22 \times 18 \times 45\frac{1}{2}$ in (55·9 × 45·7 × 115·6 cm)	
Series GC-45	High performance routine instrument. Eight models available.	Two units 55 × 16 × 20 in (139·7 × 40·6 × 50·8 cm) and 22 × 18 × 24 in (55·9 × 45·7 × 61 cm)	
Carle Instruments			
Model 8000	Low cost portable chromatograph suitable for routine control analysis and for educational purposes.	17 × 13 × 5 in (43·2 × 33 × 12·7 cm)	

GAS CHROMATOGRAPHS—*Continued*

Detectors	Specifications of basic instruments	Accessories
FID only	Specifications as for Model 1240.	As for Model 1240.
FID only	Isothermal operation from ambient to 500°. Direct on-column injection system onto glass column. Flow controller for carrier gas and for the fuel gases to dual FID.	As for Model 1240.
FID, TC or EC	Column oven temperature range from ambient to 400° with facilities for sub-ambient operation with accessory. Temperature control stable to ±0·08° and temperature meter provides direct read-out of all heated areas. Flash vaporisation and on-column injection system with gas inlet valve operable from ambient to 250°. Flow controller for carrier and auxiliary gases. 10 in recorder with disc integrator.	Gas sampling valves. Cryogenic unit for subambient operation to $-65°$.
Choice of FID, TC and EC	Isothermal operation of oven over range from 50° to 400° with a temperature stability of ±0·08° and reproducibility of ±0·1°. Flow controllers and gauges for carrier gas and auxiliary gases. On-column injection system.	Linear temperature programmer with ancillary flow controller for carrier gas. Second electrometer for dual ionisation detector operation. Flow controller and gauges for carrier and auxiliary gases.
TC	Continuously adjustable column and inlet temperature from ambient temperature to 200°. Fixed temperature detector compartment.	Gas sampling valves. Backflush and column switching valves. Carrier gas pressure regulator. Soap bubble flowmeter. Collection unit for solution IR spectroscopic examination of effluent gases. Collection unit for the preparation of KBr pellets of eluted sample for IR spectroscopic examination.

RESEARCH AND ROUTINE ANALYTICAL

Manufacturer and model	General description	Instrument dimensions (h × d × l)	Oven dimensions
Carlo Erba			
Fractovap Model GT Series 200	General purpose dual column instrument. Eight models.	25 × 30 × 16 in (64 × 76·5 × 40 cm)	9½ × 12½ × 11½ in (24 × 31·5 × 29 cm)
Fractovap Model GV Series 200	Automatic general purpose dual column instrument. Eight models.	48½ × 16½ × 19½ in (123 × 42 × 49·5 cm)	43¼ × 7¾ × 7¾ in (110 × 20 × 20 cm)
Fractovap Model GI Series 450	Single column instrument for routine biochemical analysis.	21½ × 21½ × 21½ in (55 × 55 × 55 cm)	5½ × 6 × 14½ in (14 × 15 × 37 cm)
Fractovap Model GH	Gas analysis instrument.	21½ × 21½ × 21½ in (55 × 55 × 55 cm)	6 × 5½ × 14½ in (15 × 14 × 37 cm)
Fischer and Porter Co.			
Model 310	Dual column research instrument.	18¾ × 17 × 24 in (47·6 × 43·2 × 61 cm)	
Model 400	Routine analytical instrument for biomedical use.	16¾ × 17 × 14 in (42·5 × 43·2 × 35·6 cm)	

Gas Chromatographs—*Continued*

Detectors	Specifications of basic instruments	Accessories
Choice of FID, TC, EC	Manual temperature programming from 50° to 450°. Carrier gas and auxiliary gas flow controller. Automatic solid injector. On-column injection port.	Inlet and outlet stream splitters. Automatic sample injector and automatic fraction collector. Linear temperature programmer. Carrier gas flow programmer. Pyrolyser unit. Column switching valves and gas sampling valves. Range of recorders and integrators.
Choice of FID, TC, EC	Automatic temperature programmer. Flow controller for carrier gas. Automatic solid injector and on-column injection port.	As for model GT
FID and with choice of EC, Therm., MCS	Direct on-column injection unit and automatic solid injector. Manual temperature programming and isothermal temperature control. Flow control for carrier gas and auxiliary gases.	As for model GT
He	Manual and automatic gas sampling valves. Manual temperature control for oven over the range 50° to 450°.	Linear temperature programmer. Column switching valves. Range of recorders and integrators.
FID, Ar, EC	Isothermal oven controller over range from ambient temperature to 500° with repeatability of ±0·5°. On-column sample injection. Flow control and flowmeters for carrier gas and auxiliary gases. Manual fraction collection unit.	Stream splitter to by-pass detector for effluent collection. Linear temperature programmer.
FID, Ar, EC	Specifications similar to Model 310 without flow controller and fraction collector.	Flow control and flowmeter for carrier gas and auxiliary gases.

RESEARCH AND ROUTINE ANALYTICAL

Manufacturer and model	General description	Instrument dimensions (h × d × l)	Oven dimensions
Hewlett-Packard			
Series 402	High performance dual column analytical instrument.	Oven module: 48 × 19 × 18 in (121·9 × 48·3 × 45·7 cm) Control module: 33 × 22 × 21 in (83·8 × 55·9 × 53·3 cm)	
Series 700	Low cost dual column routine instrument. Fourteen models.	Oven module: 16½ × 25 × 20½ in (41·9 × 63·5 × 52·1 cm) Bridge module: 13½ × 10 × 10½ in (34·3 × 25·4 × 26·7 cm) Electrometer module: 13½ × 10 × 10½ in (34·3 × 25·4 × 26·7 cm)	
Series 5750	Automated dual column high performance research chromatograph.	Oven module: 18 × 23 × 20 in (45·7 × 58·4 × 50·8 cm) Control module: 15 × 21 × 19 in (38·1 × 53·3 × 48·3 cm)	12 × 12 × 6 in (30·5 × 30·5 × 15·3 cm)
Series 7600A	Fully automated analytical chromatograph comprising Model 7620A chromatograph with automatic sample injection and electronic digital integrator and print-out.		

Gas Chromatographs—*Continued*

Detectors	Specifications of basic instruments	Accessories
FID	Variable oven temperature from ambient temperature to 400°. Linear temperature programmer with 12 heating rates. Pyrometer linked to five positions for temperature read-out. Injection port temperature variable up to 450°. Dual differential flow controllers keep carrier gas flow constant during programmed operation. Effluent splitter and facilities for sample collection.	^{63}Ni or ^{3}H EC detector. Additional electrometer for simultaneous operation of two ionisation detectors. Two way effluent splitter and total collection system for preparative chromatography. Choice of recorders with integrator or event markers. Electronic digital integrator.
Choice of FID, TC, EC	Choice of instruments with manual or automatic linear temperature programming operation. Isothermal operation up to 400°. Twelve heating rates from 0·5 to 30° per min up to 500° with temperature programmer.	Solid state oven temperature controller. Oven temperature programmer for isothermal instrument (12 heating rates from 0·5° to 30° per min). Flow controllers and rotameter gauges for carrier gas. On-column injection system. Gas sampling valves. Total collection system for preparative chromatography. Choice of recorders with integration. Electronic digital integrators.
TC, FID, EC, MCS	Isothermal and automatic linear temperature programmed oven operation over range 50° to 400°. Linear heating rates from 1° to 60° per min in 10 steps. Direct read-out meter for temperature in eight heated zones. Flow controller for carrier gas. Flash vaporisation injection system with minimum dead volume.	Inlet and effluent stream splitters. Gas sampling valves. Dual rotameter gauges for auxiliary gases. Electrometer suitable for simultaneous operation of two ionisation detectors. Heated collector vent for sample collection system.
FID, TC, EC, Therm.	Fully automatic isothermal and temperature programmed operation over range 70° to 500° with possibility of cryogenic operation down to $-70°$ with accessory. Ten programmed heating rates from 0·5° to 30° per min with an accuracy of $\pm 2\%$.	Automatic data reduction and computation system which translates the integrated area and retention data onto punched tape suitable for computer analysis and print-out.

RESEARCH AND ROUTINE ANALYTICAL

Manufacturer and model	General description	Instrument dimensions (h × d × l)	Oven dimensions
Perkin-Elmer Ltd.			
Model F11	High performance analytical instrument. Several models with choice of detectors and a wide variety of accessories.	Basic unit: $17 \times 18\frac{3}{4} \times 9\frac{1}{4}$ in ($43 \times 48 \times 24$ cm)	
Model 452	Robust routine single column multi-detector chromatograph.	$32 \times 16\frac{1}{2} \times 16\frac{1}{4}$ in ($81 \cdot 3 \times 41 \cdot 9 \times 41 \cdot 3$ cm)	
Model 880/881	Robust analytical dual column instrument with temperature programmer and choice of detectors.		
Model 900	High performance fully automatic temperature programmed instrument with a multi-detector system and subambient operation.	$22 \times 42 \times 20$ in ($56 \times 107 \times 51$ cm)	

GAS CHROMATOGRAPHS—*Continued*

Detectors	Specifications of basic instruments	Accessories
Choice of FID, TC, EC	Precision isothermal oven with a temperature range of 40° to 500° and a stability of 0·1° with a repeatability of 0·5°. Injection port temperature variable up to 600°.	Gas sampling valve. Glass liners for injection block. Pressure control unit for carrier gas and auxiliary gases. Flow control unit with rotometers for carrier gas. Soap bubble flowmeter. Linear pressure programming unit for carrier gas. Analyser unit for backflushing and peak cutting. Temperature readout unit. Choice of recorders.
Choice of TC, FID, EC	Manual oven temperature controller over the range 45° to 225° with a stability of ±0·1° and adjustable thermal cutout to protect column. Pressure regulator and gauge for carrier gas. Variable temperature injection block up to 350°.	Dual column valve for use with two columns in oven. Gas sampling valve. Range of recorders.
Choice of TC, FID, EC, MCS with simultaneous operation on two detectors	Linear temperature programmer from ambient temperature to 400° (12 rates up to 40° per min). Thermal cutout to protect column at 420°. Flow controller for carrier gas and auxiliary gases. Injection block temperature adjustable up to 500°.	Gas sampling valve. Inlet splitter for use with capillary column. Filter/drier assembly for gas lines. Dual flow control unit with rotometers for carrier gas. Soap bubble flowmeter. Range of recorders.
FID, TC, EC	Full automatic temperature programmer from 0° to 400° at 14 different rates with sub-ambient temperature programmes below 0° with accessory. Isothermal operation stability better than 0·1° with temperature stability of ±0·5°. Pyrometer to monitor temperature in six zones over range −100° to +400°. Dual flash vaporization or on-column injection port with adjustable temperature to 400°. Pressure controller and gauge for carrier gas.	Inlet and outlet stream splitters. Auxiliary gas pressure control unit. Soap bubble flowmeter for carrier gas. Subambient accessory for operation below 50°. Preparative accessory converts instrument into a fully automatic preparative gas chromatograph with five collection traps. Range of recorders.

RESEARCH AND ROUTINE ANALYTICAL

Manufacturer and model	General description	Instrument dimensions (h × d × l)	Oven dimensions
Model RGC-170	Radio-gas chromatograph for rapid determination of ^{14}C and ^{3}H.		
Phase Separations Ltd.			
Model LC-2	Low cost instrument suitable for gas analysis and educational purposes.		11 × 8 × 7 in (27·9 × 20·3 × 29·2 cm)
Pye-Unicam Ltd.			
Series 104	High performance analytical instrument. Fourteen models with choice of detector and temperature programmer.	Dimensions vary with model.	10 × 10 × 4 in (25·4 × 25·4 × 10·2 cm)
Series 106	Fully automated analytical chromatograph with automatic solids injection unit. Basic unit is model 4 of 104 series.		10 × 10 × 4 in (25·4 × 25·4 × 10·2 cm)
Tracor			
Series MT-150	Modular high performance system for special applications.		

Gas Chromatographs—*Continued*

Detectors	Specifications of basic instruments	Accessories
Proportional continuous flow counter tube	Continuous flow reaction tube converts the compounds by hydrogenative cracking or by oxidative degradation (for ^{14}C) into a form suitable for analysis. Instrument permits direct insertion of single samples or, via the gas chromatograph, the analysis of complex mixtures of radioactive compounds.	
TC with optional choice of GD and FID	Isothermal oven operation over range 45° to 250° with temperature stability of $\pm 0.1°$ and repeatability of better than 1°. On-column and flash evaporation injection facilities with injection temperature variable over range from ambient to 500°.	
Choice of FID, TC, EC	Isothermal oven operation up to 500° with a stability of $\pm 0.1°$. Thermal fuses for column protection at 420° and 570°. Detector supported in column oven. Temperature programmed model also available and carrier gas flow control on certain models.	Gas flow control units for carrier gas purifying unit. Detector oven temperature controller. Temperature programmer controller. Manual preparative unit. Pyrolysis probe. Gas sampling valves.
FID	Basic model 104–4 with automatic solids injection unit. (This unit can be purchased separately and fitted to any Series 104 chromatograph.)	As for 104 series. TC and EC detectors.
FID, TC, EC, Photo., US	Air blower oven for isothermal or programmed operation and, with accessory, for subambient operation.	Cryothermal system for subambient operation to $-180°$. Digital integrator and range of recorders. Gas sampling valves.

RESEARCH AND ROUTINE ANALYTICAL

Manufacturer and model	General description	Instrument dimensions (h × d × l)	Oven dimensions
Series MT-160	Isothermal and temperature programmed models suitable for routine analytical work.	16 × 16 × 34 in (40·6 × 40·6 × 86·4 cm)	
Series MT-220	High performance dual column research chromatograph. Four basic models with choice of detector.	56 × 17 × 33 in (142·2 × 43·2 × 83·8 cm)	1100 cu in
Series GC-2000R	Dual Column Modular chromatograph for analytical research and product control.	30 × 15 × 35 in (76·2 × 38·1 × 88·9 cm)	650 cu in

Varian
Aerograph HY-FI III Series 1200	Single column precision routine instrument.	15 × 15¾ × 15¼ in (38·1 × 40 × 38·7 cm)	6 × 9 × 9 in (15 × 23 × 23 cm)

Gas Chromatographs—*Continued*

Detectors	Specifications of basic instruments	Accessories
TC, FID, EC, Photo	Isothermal model includes manual temperature programmer to 500° with direct read-out pyrometer monitor. Model MT 160DP fitted with automatic Temperature programmer with 49 discrete heating rates from 1° to 50° per min.	The Micro Universal inlet system permits the use of a range of accessories for the introduction of liquids, solids and gases. Glass conversion kit for sample introduction for biomedical work. Digital integrator and range of recorders.
Choice of TC, FID, EC, Photo., Therm.	Isothermal oven temperature controller reproducible to within 0·5°. Temperature range from ambient to 500°. Temperature programmed models have 49 discrete heating rates from 1° to 50° per min. Electronic timers reproducible to within 0·6 sec. On-column injection system and facilities for flash vaporisation in glass lined injection port.	
Choice of FID, TC, EC, Therm. US, A	Isothermal and temperature programmed operation up to 500°. Programmed heating rates from 0·5° to 50° in 12 steps with a temperature stability of ±0·3°. Flash vaporisation injection system with provision for on-column injection. Facilities for gas sampling valves.	
FID, EC, Therm.	Oven temperature range from ambient to 400° with direct reading pyrometer monitor. Flash vaporisation and on-column injection system with variable temperature from ambient to 400° Model 1200–2: Isothermal temperature controller accurate to ±0·5°. Model 1200–1: Linear temperature programmer with 10 heating rates from 0·5° to 20° per min.	Gas sampling valves. Column switching valves and backflush unit. Micro-sample collection unit. Pyrolyser unit. Auxiliary gas flow controller. Rotometer for carrier gas. Range of recorders and digital integrators. Hydrogen generator.

G

RESEARCH AND ROUTINE ANALYTICAL

Manufacturer and model	General description	Instruments dimensions (h × d × l)	Oven dimensions
Moduline Series 1700 and 1800	High precision research chromatograph. Twenty-four models.	Series 1700: $20\frac{1}{4} \times 21\frac{1}{2} \times 19\frac{3}{8}$ in (51·4 × 54·6 × 49·2 cm) Series 1800: $20\frac{1}{4} \times 21\frac{1}{2} \times 29\frac{1}{2}$ in (51·4 × 54·6 × 74·9 cm)	8 × 13 × 10 in (20 × 33 × 27 cm)
Series 2100	Routine instrument of use in biomedical applications. Four column operation with multi-detector operation.	67 × 24 × 37 in (170 × 61 × 94 cm)	38 × 11 × 7 in (97 × 28 × 18 cm)
Model 90-P3	All purpose single column instrument.	$18\frac{1}{2} \times 14 \times 16\frac{3}{8}$ in (48 × 36 × 42 cm)	$8 \times 7\frac{1}{2} \times 6\frac{1}{2}$ in (20·3 × 19 × 16·5 cm)
Victoreen Series 4000	Twenty-two unimodular instruments with choice of detector and cryogenic operation. Suitable for routine or research usage. Four column operation.	Basic system: $23 \times 18 \times 25\frac{3}{4}$ in (58·4 × 45·7 × 65·4 cm)	$14 \times 10 \times 9\frac{1}{2}$ in (35·6 × 25·4 × 24·1 cm)

Gas Chromatographs—*Continued*

Detectors	Specifications of basic instruments	Accessories
TC, FID, EC, Therm.	Isothermal and automatic linear temperature programming operation from ambient to 400° with direct reading temperature monitor and adjustable (150–400°) thermal fuse to prevent overheating of column. Ten programmed heating rates from 1° to 40° per min with stability of $\pm 0.5°$.	Gas sampling valves. Inlet and outlet stream splitters. Backflush unit. Manual collection unit. Dual flow controller for auxiliary gases. Range of recorders and digital integrators. Hydrogen generator.
FID, EC, Therm.	Isothermal and linear temperature programming operation from ambient to 400° with reproducibility of $\pm 1°$ and stability of $\pm 0.3°$ and temperature limiting device to protect column. Ten programmed heating rates and direct reading temperature monitor. On-column and flash vaporisation injection system with variable temperature from ambient to 400° reproducible to $\pm 2°$ and a $\pm 0.5°$ stability. Independent differential flow controllers for carrier gas.	Dual flow controller for auxiliary gases. Range of recorders and digital integrators. Hydrogen generator.
TC	Oven temperature range from ambient to 400° with adjustable thermal fuse to protect column. Isothermal and non-linear manual temperature programming control with four position pyrometer monitor. Flash vaporisation and on-column injection system with variable temperature up to 400°.	Glass liners for injection block. Isothermal proportional controller. Linear temperature programmer. Six-way gas sampling valve. Backflush and column switching unit. Manual collection unit.
TC, FID, EC	Basic instrument: Manual temperature controller from ambient to 500°. Temperature programmer available with 16 heating rates reproducible to $\pm 0.5°$ from 0.5° to 60° per min. Cryogenic operation to $-195°$ possible with accessory.	Inlet and outlet stream splitters. Pyrolysis unit. Cryogenic unit for subambient operation. Digital logarithmic electrometer. Integrator with printout facilities.

PREPARATIVE GAS

Manufacturer and Model	Control System	Injection System
Carlo Erba		
Fractovap Model P Series 1300	Four timers for temperature cycle adjustable from 6 min to 6 hr. Linear temperature programming from 0·5 to 20°/min over range ambient to 300°. Automatic cool down period.	Automatic pneumatic injector operated by motor. Two sample volume range variable from 0·02 to 2·5 ml and from 0·1 to 12 ml. Injection temperature variable from 50 to 350°.
Hewlett-Packard		
Model 775 (automatic)	Ten temperature programming rates from 0·25 to 10°/min for manual or automatic operation from ambient temperature to 325°. Adjustable times for pre- and post-programme periods from 0 to 60 min. Automatic cool down cycle.	Piston injection system powered by carrier gas delivering from 0·25 to 12 ml from 300 ml reservoir. Adjustable timed injector variable from 0 to 50 sec to deliver up to 125 ml. Manual injection system via separate port. Optional reservoirs of 1000 and 2250 ml capacity. Flash vaporiser temperature variable to 350°.
Model 776 (manual)	Isothermal operation to 300°. Optional temperature programmer providing heating rates from 0·5 to 30°/min.	Injection by means of a pair of toggle valves applying carrier gas pressure to sample reservoir (75 ml capacity). Injected volume controlled by dip tube. Separate analytical injection port. Flash vaporiser temperature variable to 350°.

CHROMATOGRAPHS

Column System	Collection System	Overall Size, Detector System and General Specifications
Oven size: 45 × 8 × 7 in (114 × 20 × 18 cm) for dual column operation. Upper temperature limit 300°.	Five traps collect preselected fractions detected by peak height discriminator (optional extra). Pneumatic valve controlling collection located after traps and fitted with a reverse purge gas system to clear delivery tubes. Temperature of delivery tubes maintained between ambient and 300°. Traps cooled in individual Dewer flasks.	38 × 23 × 65 in (98 × 60 × 165 cm) TC detector (optional FID and EC detector with stream splitter). Optional carrier gas recycle unit for use with helium.
Oven size: 12 × 6½ × 53 in (30·5 × 16·5 × 134·6 cm) accommodates up to 400 ft × ⅜ in or 80 in × 4 in column. Upper temperature limit 325°. Variable high temperature safety cut off.	Automatic collection of up to six components from a 17 component system. Variable temperature control up to 350° for the seven outlet ports. 50 ml glass spiral traps cooled in stainless steel bath by an optional refrigeration unit. Larger collection traps as optional extra.	57 × 19 × 60 in (144·8 × 48·3 × 152·4 cm) TC detector. Separate rotometer and differential flow controllers for carrier gas to preparative and analytical columns. Rotometer and needle valve for auxiliary gases for optional FID. Five position direct read out temperature monitor.
Oven size: 12 × 7 × 48 in (30·5 × 17·8 × 121·9 cm) accommodating up to 400 ft × ⅜ in or 80 in × 4 in column. Optional auxiliary oven double capacity. Upper temperature limit 300°. High temperature safety cut off adjustable to 350°.	Five position heated manifold which accepts 50 ml aluminium traps. 2, 10 and 50 ml glass traps as optional extra. 1000 ml stainless steel trap also available.	48 × 23 × 63 in (121·9 × 58·4 × 160 cm) FID with variable stream splitter. Separate rotometer and needle valves for carrier gas to preparative and analytical columns. Needle valve for auxiliary gases for FID.

G§

Preparative Gas

Manufacturer and Model	Control System	Injection System
Hupe		
Model APG 402	Fully transistorised system. Time controlled initial isothermal period, temperature programme and post programme periods. Cool down cycle controlled by column temperature measurements. Six temperature programme rates from 0·25 to 8°/min.	Electropneumatic injection system. 60, 100 or 200 ml reservoir. Sample size ranges: 0·1 to 2 ml, 0·5 to 6 ml, 1 to 10 ml. Injection chamber temperature variable from ambient to 400°.
Nester-Faust		
Model 850 Prepkromatic	Five linear heating rates from 2 to 10°/min from ambient to 500°. Automatic cooling cycle. Cycle comprises three separate 0 to 30 min isothermal periods and two separate 0 to 30 min programmed periods.	Positive displacement motor driven syringe type pump injection unit. Sample sizes from 0·2 to 2·2 ml. Total reservoir capacity 40 ml. Optional macro system with 350 ml reservoir and sample sizes from 2 to 22 ml. Three separate flash vaporisation channels with temperature range from ambient to 500°. Manual injection possible.
Model 750 Prepkromatic	Variable temperature programme heating rates from 1 to 20°/min. From ambient to 500°. Automatic cool down cycle. Programmed cycle comprises two separate isothermal periods and one temperature programmed period.	Injection system as for model 850. Injection volumes variable from 0·2 to 2 ml from 40 ml reservoir.

Chromatographs—*Continued*

Column System	Collection System	Overall Size, Detector System and General Specifications
Oven size: 33 × 8 × 43 in (83·8 × 20·3 × 109·2 cm) capable of heating up to 32 in × 1 in or 150 ft × $\frac{3}{8}$ in columns. Upper temperature limit of 325°.	Five fraction collection traps and one waste trap. Temperature of outlets controlled from ambient to 400°. Traps cooled in individual Dewer flasks. Solenoid valves controlling the collection unit located after the traps and fitted with a reverse purge gas system to prevent cross contamination of fractions. Delay switches electronically controlled to permit the bypass of up to eight components.	48 × 23$\frac{1}{2}$ × 59 in (121·9 × 59·7 × 149·9 cm) FID with variable stream splitter.
Oven size: 41 × 13 × 10 in (104·1 × 33 × 25·4 cm) accommodating 6 ft × $\frac{3}{4}$ in "biwall" columns. Upper temperature limit of 500°.	Motor driven collection valves delivering to six glass collection traps and one common waste trap. All traps cooled in individual Dewer flasks. Temperature of outlet ports adjustable from ambient to 500°. Collection unit activated by switch on recorder or by an optional peak selector switch.	16 × 16$\frac{1}{2}$ × 45 in (40·6 × 41·9 × 115·6 cm) with a control console 21 × 19 × 25 in (55·2 × 48·3 × 64·8 cm) TC detector. FID optional extra. Two metering valves for carrier gas flow for column and reference detector lines. Five position direct read out temperature monitor. Manual over ride system for injection and fraction collection.
Oven size: 35 × 6$\frac{1}{2}$ × 7 in (88·9 × 16·5 × 17·8 cm) accommodating 30 in × $\frac{3}{4}$ in "biwall" column. Upper temperature limit of 500°.	Collection system as for model 850.	14 × 24$\frac{1}{2}$ × 64 in (35·6 × 62·2 × 162·6 cm). TC detector. Five position direct read-out temperature monitor. Manual system switch deactivates all automatic systems.

PREPARATIVE GAS

Manufacturer and Model	Control System	Injection System
Perkin Elmer Model F21	Fully transistorised control. Time controlled programmes for warm up period (1 to 31 min), upper temperature period (1 to 31 min) and cool down cycle (1 to 31 min). Exponential temperature programmer over range 50 to 350°.	Electro-pneumatic operated injection system controlled by carrier gas pressure. Three different sizes of dosing capillary. Timed sample injection to deliver from 10 μl to 3·5 ml. Four sizes of reservoir (2, 25, 50 and 100 ml). Injection block temperature variable from 100 to 400°. Two different vaporising capillaries.
Pye-Unicam Series 105	Model 5. Isothermal instrument. Model 15. Temperature programmed instrument. Peak selection and temperature programme controlled by a continuous punched tape or by an auto-indexing unit. Twelve linear programmed heating rates from 0·5 to 24°/min from ambient to 500°. Automatic cool down to initial temperature. Manual operation possible.	Pneumatic displacement of sample from 50 ml reservoir by carrier gas controlled by switching valves. Sample size controlled by carrier gas pressure regulator to reservoir. Minimum volume 100 μl. Carrier gas clears delivery capillary after each injection. Manual injection possible.
Varian Aerograph Autoprep Model 712	Linear temperature programmer with ten heating rates from 0·5 to 20°/min. Variable time control for pre- and post-collection periods (up to 3 hr). Variable delay in temperature programmed cycle up to 3 hr. Cool down cycle variable up to 15 min.	Automatic pneumatic injection pump delivers from 0·1 to 6 ml sample. Injection block temperature variable to 400°.

CHROMATOGRAPHS—*Continued*

Column System	Collection System	Overall Size, Detector System and General Specifications
Oven accommodates up to 14¾ ft × ⅜ in or 5¼ ft × 1 in columns. Upper temperature limit of 350°.	Collection manifold controlled by solenoid valves located after traps. Seven components may be collected from 16 component system with common waste trap. Threshold control variable over measuring range from 5 to 95%. Traps cooled in two Dewer flasks. Push-button operation for manual collection.	37¾ × 15¾ × 29½ in (96 × 40 × 75 cm) FID with variable size stream splitter controls collection manifold. Optional linear temperature programmer. Optional individual threshold controls for each component. Optional flow controls for auxiliary gases.
10 × 10 × 4 in (25·4 × 25·4 × 10·2 cm). Upper temperature limit 500° with stability of > 0·1°.	Twelve components can be collected. Twenty-four cooled traps are accommodated on the collection turntable and the outlet from the column terminates in a needle which is pneumatically actuated to pierce the rubber caps on the traps during the collection of selected components. Effluent can be diverted to waste at any interval.	FID fitted with 1 : 100 stream splitter. Optional extras include a refrigeration unit for sub-ambient operation. All accessories available for the Series 104 analytical chromatograph can be used with the series 105 instrument. The Model 5 instrument can be converted for temperature programming with an automatic programmer controller.
Cylindrical oven — 18 in diameter × 8 in accommodating up to 250 ft × ⅜ in or 100 ft × ¾ in columns. Upper temperature limit of 350°.	Automatic collection of up to eight components with waste position between each collection.	48⅛ × 30 × 19⅞ in (122·5 × 76·2 × 50·5 cm) FID with variable effluent splitter. Soap bubble flow meter for carrier gas. Direct reading temperature monitor for column. Optional hydrogen generator. Optional refrigerated bath for traps operable to −50°.

COMPANY DIRECTORY

A—Analytical and precision chromatographs: P—Preparative scale chromatographs: CR—Cryogenic chromatographs: PY—Chromatographs with pyrolysis units: PR—Process control chromatographs and custom built industrial units: TP—Temperature programme accessories: FP—Flow programme accessories: CS—Column and stationary phases: I—Integration units (digital or chart): MS—Combined mass spectrometry—gas chromatography units.

Company									
Amscor PO Box HH Angleton Texas 77515, USA	A				PR	TP			
Barber-Colman Co. Chromatography Products Division 1300 Rock Street Rockford Illinois 61101, USA	A	P	CR	PY		TP		CS	I
Becker Delft N.V. Vulcanusweg 113 PO Box 219 Delft, Holland	A	P		PY	PR	TP	FP	CS	I
Beckman Instruments Inc. 2500 Harbor Boulevard Fullerton California 92634, USA	A		CR		PR	TP		CS	I
Beckman Instruments Ltd. Glenrothes Fife, Scotland	A		CR		PR	TP		CS	I
The Bendix Corp. Process Instruments Division PO Drawer 477 Ronceverte W. Virginia 24970, USA	A	P	CR		PR	TP		CS	MS
Carle Instruments Inc. 1141 East Ash Avenue Fullerton California 92631, USA	A								

Carlo Erba S.A. Scientific Instrument Division Via Carlo Imbonati 24 Milano, Italy	A	P	CR	PY	PR	TP	FP	CS	I
Fisher and Porter Co. Lab-Crest Scientific Division County Line Road Warminster Pennsylvania 18974, USA	A	P	CR			TP		CS	
Fisher Scientific Co. 711 Forbes Avenue Pittsburgh Pennsylvania 15219, USA	A		CR			TP		CS	I
Fisons Scientific Apparatus Ltd. Loughborough Leics, UK	A								I
Glowall Corp. Easton and Davisville Roads Willow Grove Pennsylvania 19090, USA	A		CR			TP		CS		MS
Gow-Mac Instrument Co. 100 Kings Road Madison New York 07940, USA	A								
HCL Scientific Inc. 1800 Broadway PO Box 4223 Rockford Illinois 61110, USA	A		CR			TP		CS	I
Hewlett-Packard Co. Avondale Division Route 41 Avondale Pennsylvania 19311, USA	A	P	CR	PY	PR	TP		CS	I
Hewlett-Packard Ltd. 224 Bath Road Slough Bucks, UK	A	P	CR	PY	PR	TP		CS	I
LKB Instruments Inc. 12221 Parklawn Drive Rockville Maryland 20852, USA										MS

Loenco Inc. 1062 Linda Vista Avenue Mt. View California 94040, USA	A	P	CR	PY	PR	TP		CS	I	MS
Mine Safety Appliances Co. Instrument Division 201 North Braddock Avenue St. Louis Missouri 63103, USA	A		CR		PR	TP		CS	I	
Nester/Faust Mfg. Corp. 2401 Ogletown Road Newark Delaware 19711, USA		P						CS	I	
Perkin-Elmer Corp. 702-G Main Avenue Norwalk Connecticut 06852, USA	A	P	CR	PY	PR	TP	FP	CS	I	MS
Perkin-Elmer Ltd. Post Office Lane Beaconsfield Bucks, UK	A	P	CR	PY	PR	TP	FP	CS	I	
Phase Separations Ltd. Deeside Industrial Estate Queensferry Flintshire, UK	A	P		PY	PR		FP	CS	I	
Philips Electronic Instruments 750 South Fulton Avenue Mt. Vernon New York 10550, USA	A	P		PY	PR	TP		CS		
Pye-Unicam Ltd. York Street Cambridge, UK	A	P		PY	PR	TP		CS		
Servomex Controls Ltd. Crowborough Sussex, UK	A				PR	TP	FP	CS	I	
Siemans AG Rheinbruckenstrasse 50 75 Karlsruhe 21 W. Germany	A	P	CR		PR	TP		CS	I	

Techmation Ltd. 58 Edgware Way Edgware Middlesex, UK	A	P	CR			TP		I	
Thermco Instrument Corp. PO Box 309 LaPorte Indiana 46350, USA	A			PY				CS	
Tracor Inc. Analytical Division 6500 Tracor Lane Austin Texas 78721, USA	A	P	CR	PY		TP		CS	I MS
Varian Aerograph 2700 Mitchell Drive Walnut Creek California 94598, USA	A	P	CR	PY		TP	FP	CS	I MS
Varian Aerograph Haletop Civic Centre Wythenshawe Manchester 22, UK	A	P	CR	PY		TP	FP	CS	I MS
Victoreen Instruments Fisher Scientific Co. 711 Forbes Avenue Pittsburgh Pennsylvania 15219, USA	A	P	CR	PY	PR	TP		CS	I

AUTHOR INDEX

Numbers in parentheses are reference numbers and are included to assist in locating references. Numbers in italics refer to the page on which the reference is listed.

A

Advances in Chromatography 1965, 3 (14), *19*
Ahrens, E. H., 70 (8, 11), *75*
Alport, N. L., 103 (28), *115*
Anavaer, B. I., 60 (4), *74*
Annis, J. L., 109 (40), *115*
Ardnt, F., 73 (31) *75*
Arens, C. L., 163 (6), *165*
Atkinson, E. P., 32 (6), *55*
Atkinson, E. R., 163 (11), *165*

B

Barr, J. K., 84 (7), 88 (7), *115*
Bartlet, J. C., 82 (5), *114*
Bartz, A. M., 103 (26), 104 (26), *115*
Becker, E. W., 111 (48), *116*
Beersum, W. van, 34 (7), *55*
Belleb, K., 37 (13), *55*
Biemann, K., 109 (45, 47), *116*
Birkofer, L., 71 (27), *75*
Black, D. R., 112 (50), *116*
Bobbitt, J. M., 97 (19), *115*
Boucher, E. A., 70 (15), 72 (15), *75*
Brace, R. O., 126 (10), *133*
Brody, S. S., 52 (28), *56*
Brooks, C. J. W., 70 (24), *75*
Brown, I., 61 (5), *75*, 93 (10), *115*
Brown, R. A., 103 (27), 104 (27), *115*
Bruner, F. A., 37 (11), *55*
Bryce, W. A., 41 (19), *55*
Burchfield, H. P., 18 (27), *20*

C

Cansisius College Gas Chromatography Institute 1961, 3 (11), *19*
Capella, P., 70 (22), 71 (22), *75*
Caron, B. C., 83 (6), *114*
Cartoni, G. P., 37 (11), *55*
Cassanova, J., 163 (8), *165*
Cassidy, H. G., 1 (1), *18*
Chaney, J. E., 52 (28), *56*
Charles-Sigler, R., 163 (5), 164 (5, 14, 21, 22, 23), *165*
Chromatographic Reviews 1959, 3 (15), *19*
Chu, F., 70 (8), *75*
Clayton, J. D., 81 (4), *114*
Corey, E. J., 163 (8), *165*
Cornwell, D. G., 74 (38), *75*
Cort, L. A., 163 (6), *165*
Cruickshank, P. A., 71 (25), *75*

D

Dal Nogare, S., 18 (23), *19*
Day, E. A., 109 (43), *116*
Deans, D. R., 126 (12), *133*
Desty, D. H., 35 (8), *55*
Dewar, R. A., 45 (24), *56*
Doering, W. E., 163 (2), *165*
Dorsey, J. A., 109 (46), *116*
Dvoretzky, I., 163 (4), *165*
Dwyer, J. L., 99 (23), 101 (23), 102 (23), *115*

E

Ebert, A. A., 109 (38), *115*
Ehrler, A. J., 123 (6), *132*
Elliot, W. H., 114 (53), *116*
Elsken, R. H., 105 (31), 106 (31), 107 (35), *115*
Ettre, L. S., 18 (26), *20*, 35 (9), 37 (13), *55*, 60 (1), *74*

F

Feibush, B., 164 (21, 22, 23), *165*
Feigl, F., 95 (15), *115*
Feit, E. D., 96 (17), *115*
Felkin, H., 163 (7), *165*
Finkbeiner, H., 70 (13), *75*
Fischer, G., 163 (5), 164 (5), *165*

Flath, R. A., 105 (31), 106 (31, 34), 107 (35), 112 (50), *115*, *116*
Flett, M. St. C., 103 (25), *115*
Folmer, O. F., 70 (16), *75*
Fowlis, I. A., 104 (30), 114 (30), *115*
Freeman, S. K., 103 (29), 104 (29), *115*
Friedmann, S., 72 (29), *75*
Fries, I., 114 (53), *116*

G

Gardiner, W. L., 70 (18, 23), 71 (23, 28), *75*
Gas Chromatography Abstracts 1958–1962, 3 (8), *18*
Gault, Y., 163 (7), *165*
Gehrke, C. W., 73 (32), *75*
Gesser, H. D., 120 (2), *132*
Giesecke, W., 71 (26), 73 (26), *75*
Gil-Av, E., 163 (5), 164 (5, 14, 21, 22, 23), *165*
Giuffrida, L., 46 (27), *56*
Glenn, T. H., 42 (20), *55*
Goering, H. L., 163 (12), *165*
Goerlitz, D. F., 73 (32), *75*
Gohlke, R. S., 109 (37), *115*
Goldfein, A., 26 (1), *55*

H

Habgood, H. W., 117 (1), *132*
Halasz, I., 37 (12, 14, 15), *55*
Halpern, B., 163 (3), 164 (18, 19, 20), *165*
Hamilton, W. C., 79 (2), *114*
Harley, J., 45 (23), *56*
Harris, W. E., 117 (1), *132*
Haslam, J., 99 (22), 100 (22), *115*
Hawes, J. E., 112 (51), *116*
Hawkes, S. J., 89 (9), *115*
Hedgley, E. J., 70 (20), *75*
Heine, E., 37 (14, 15), *55*
Henderson, N., 106 (34), 107 (35), *115*
Hildebrand, G. P., 64 (7), *75*
Hill, H. C., 113 (52), *116*
Hoff, J. E., 96 (17), *115*
Holmes, J. C., 109 (41), *115*
Horning, E. C., 70 (14, 15, 18, 22, 23, 24), 71 (22, 23, 28), 72 (14, 15), *75*
Horning, M. G., 70 (14, 15), 72 (14, 15), *75*
Horrocks, L. A., 74 (38), *75*
Horvath, C., 37 (12), *55*
Hougen, F. W., 120 (2), *132*
Howard, T. J., 163 (6), *165*
Hughes, J., 103 (25), *115*
Hunt, R. H., 109 (46), *116*
Huyton, F. H., 34 (7), *55*
Hyder, G., 3 (18), *19*

I

Issenberg, P., 123 (7), *133*

J

James, A. T., 2 (3, 4, 5, 6), *18*, 39 (16, 17), *55*
Janek, J., 42 (21), 45 (21), *56*
Jeffrey, P. G., 28 (5), *55*
Jeffs, A. R., 99 (22), 100 (22), *115*
Jennings, F. C., 163 (11), *165*
Johns, T., 126 (10), *133*
Johnson, R. D., 164 (16), *165*
Jones, C. N., 123 (5), *132*
Jones, H. G., 70 (21), *75*
Jones, H. J., 85 (8), *115*
Juvet, R. S., 18 (23), *19*

K

Kaiser, R., 11 (20), *19*, 96 (18), *115*
Karger, B. L., 163 (9), 164 (13, 19), *165*
Karmen, A., 45 (22), 46 (26, 27), *56*
Kates, M., 74 (37), *75*
Kaufman, M. L., 72 (29), *75*
Kaufmann, H. P., 73 (34), 74 (34), *75*
Keane, W. J., 164 (13), *165*
Kelker, H., 163 (1), *165*
Keulemans, A. I. M., 18 (24), *19*
Kipping, P. J., 28 (5), *55*
Klebe, J. F., 70 (13), *75*
Konig, W., 164 (17), *165*
Kovats, E., 93 (11, 12), 94 (13), 95 (13), *115*
Kratz, P. D., 123 (3), *132*
Krejci, M., 42 (21), 45 (21), *56*
Kurischka, K., 26 (2), *55*

L

Langer, S. H., 70 (12), 71 (12), *75*
Larson, D. W., 163 (12), *165*

Lee, F. A., 70 (10), *75*
Leftault, C. J., 37 (10), *55*
Levy, R. L., 27 (4), *55*, 120 (2), *132*
Lewis, J. S., 3 (16), *19*
Lindeman, L. P., 109 (40), *115*
Link, W. E., 70 (17), 74 (17), *75*
Litovtseva, I. I., 60 (4), *74*
Littlewood, A. B., 18 (21), *19*, 60 (3), *74*, 99 (21), 100 (21), 108 (36), *115*, 126 (11), *133*
Lively, L. D., 79 (2), *114*
Loe, L. B., 163 (6), *165*
Low, M. J. D., 103 (29), 104 (29). *115*
Lundin, R. E., 105 (31, 32), 106 (31, 34), 107 (35), 108 (32), *115*

M

McComas, D. M., 26 (1), *55*
McFadden, W. H., 63 (6), *75*, 105 (32), 108 (32), 109 (43), *115*, *116*
McInnes, A. G., 74 (37), *75*
McReynolds, W. O., 3 (17), *19*
McWilliam, I. G., 45 (24), *56*
Maczka, H., 123 (6), *132*
Mallaby, R., 112 (51), *116*
Mankel, G., 73 (34), 74 (34), *75*
Mason, M. E., 74 (35), *75*
Martin, A. E., 99 (24), 102 (24), 104 (24), *115*
Martin, A. J. P., 2 (2, 3, 4, 5), *18*, 39 (16, 17), *55*
Mattick, L. R., 70 (10), *75*
Mazor, L., 124 (9), *133*
Mefferd, R. B., 81 (4), *114*
Merritt, C., 95 (14), *115*, 123 (4), *132*, 123 (7), *133*
Metcalfe, L. D., 70 (9), *75*
Miller, D. O., 109 (39), *115*
Mon, T. R., 107 (35), *115*
Morrell, F. A., 109 (41), *115*
Morrissette, R. A., 70 (17), 74 (17), *75*
Moss, A. M., 70 (14, 15), 72 (14, 15), *75*,

N

Neely, W. B., 72 (30), *75*
Nel, W., 45 (23), *56*
Nelson, F. A., 105 (33), *115*
Nerheim, A. G., 26 (3), *55*

Nott, J., 72 (30), *75*
Nurok, D., 163 (5, 10), 164 (5, 14), *165*

O

Oette, K., 70 (11), *75*
O'Neal, M. J., 109 (46), *116*
Ongkiehong, L., 46 (25), *56*, 79 (1), *114*
Overend, W. G., 70 (20), *75*
Oyama, V. I., 164 (15, 16), *165*

P

Pantages, P., 70 (12), 71 (12), *75*
Patton, S., 73 (33), *75*
Pederson, C. S., 70 (10), *75*
Perkins, G., 70 (16), *75*, 79 (2), *114*
Perry, M. B., 70 (21), *75*
Petitjean, D. L., 37 (10), *55*
Pollock, G. E., 164 (15, 16), *165*
Preliminary Recommendations on Nomenclature and Presentation of Data in Gas Chromatography, 1960, 11 (19), 17 (19), *19*
Prescott, B. O., 55 (29), *56*
Preston, S. T., 3 (18), *19*
Pretorius, V., 45 (23), *56*
Prox, A., 164 (17), *165*
Purcell, J. E., 37 (13), *55*
Purnell, J. H., 18 (22), *19*

R

Ray, N. H., 80 (3), *114*
Reed, R. I., 113 (52), *116*
Reilly, C. N., 64 (7), *75*
Richardson, D. B., 163, (4), *165*
Rijnders, G. W. A., 34 (7), *55*
Ritter, A., 71 (27), *75*
Robert-Lopes, M. T., 113 (52), *116*
Roberts, C. B., 72 (30), *75*
Robertson, D. H., 123 (7), *133*
Rohr, W., 70 (19), *75*
Rohrschneider, L., 60 (2), *74*
Ronayheb, G. M., 79 (2), *114*
Rose, H. C., 163 (9), 164 (13), *165*
Roth, W. R., 163 (2), *165*
Ruhl, H. D., 103 (26), 104 (26), *115*
Russell, C. P., 89 (9), *115*

Ryce, S. A., 41 (19), *55*
Ryhage, R., 109 (42), 110 (42), 111 (49), 114 (53), *116*

S

Sakharov, V. M., 60 (4), *74*
Sawyer, D. T., 84 (7), 88 (7), *115*
Scherer, J. R., 105 (32), 108 (32), *115*
Schivizhoffen, E. von, 163 (1), *165*
Schmidhammer, L., 164 (17), *165*
Schmitz, A. A., 70 (9), *75*
Schneider, F. L., 95 (16), *115*
Schupp, O. E., 18 (25), *20*
Scott, R. P. W., 40 (18), *55*, 104 (30), 114 (30), *115*, 124 (8), *133*
Sheehan, J. C., 71 (25), *75*
Simmons, M. C., 163 (4), *165*
Smith, D. M., 82 (5), *114*
Smith, G. H., 2 (5), *18*
Snyder, F., 74 (36), *75*
Staab, H. A., 70 (19), *75*
Stenhagen, E. Z., 109 (44), *116*
Stephen, A. M., 163 (10), *165*
Stern, R., 163 (11), *165*
Stern, R. L., 163 (9), 164 (13), *165*
Stoffel, W., 70 (8), *75*
Storrs, E. E., 18 (27), *20*
Summers, R. M., 81 (4), *114*
Svojanovsky, V., 42 (21), 45 (21), *56*
Sweeley, C. C., 114 (53), *116*
Symposia of East German Gas Chromatography Group 1958, 3 (12), *19*
Symposia of Gams, 1961, 3 (13), *19*
Symposia of Gas Chromatography, 1956, 3 (7), *18*
Symposia of Instrument Society of America 1947, 3 (9), *18*
Symposia on Advances in Gas Chromatography 1963, 3 (10), *19*
Synge, R. L. M., 2 (2), *18*

T

Takacs, J., 124 (9), *133*
Tattrie, N. H., 74 (37), *75*
Taylor, G. L., 163 (10), *165*

Teranishi, R., 105 (31, 32), 106 (31, 34), 107 (35), 108 (32), 112 (50), *115, 116*
Tesarik, K., 42 (21), 45 (21), *56*
Thin Layer Chromatography, 1965, 97 (20), *115*
Thomas, P. J., 99 (23), 101 (23), 102 (23), *115*
Tuey, G. A. P., 32 (6), *55*
Turkel'taub, N. M., 60 (4), *74*

V

Vandenberg, J., 26 (2), *55*
Van den Dool, H., 123 (3), *132*
Van den Heuval, W. J. A., 70 (18, 23), 71 (23), *75*
Van der Vlies, C., 83 (6), *114*
Villalobos, R., 126 (10), *133*
Von Rühlmann, K., 71 (26), 73 (26), *75*
Vorbeck, M. L., 70 (10), *75*

W

Waller, G. R., 74 (35), *75*
Walsh, J. T., 95 (14), *115*, 123 (4), *132*
Ward, W. M., 103 (28), *115*
Watson, J. T., 109 (45, 47), *116*
Wehrli, A., 94 (13), 95 (13), *115*
Weinstein, B., 163 (3), *165*
Welti, D., 104 (30), 114 (30), *115*
Westley, J. W., 163 (3), 164 (18, 19, 20), *165*
Weygand, F., 164 (17), *165*
White, D. M., 70 (13), *75*
White, J. U., 103 (28), *115*
Wilkins, T., 104 (30), 114 (30), *115*
Wilks, P. A., 103 (27), 104 (27), *115*
Williams, V. P., 112 (51), *116*
Willis, H. A., 99 (22), 100 (22), *115*
Winefordner, J. D., 42 (20), *55*
Wise, H. L., 55 (29), *56*
Wood, R., 74 (36), *75*

Z

Zhuhkovitskii, A. A., 60 (4), *74*
Zlatkis, A., 18 (26), *20*

SUBJECT INDEX

A

Acylation of sample, procedures for, 74
Adjusted retention volume *see* Retention Volume, adjusted
Alcohols, derivatives of, 70–74
  detection of, 96
  separation of, 160, 161
Aldehydes, detection of, 96
  separation of, 159, 160
Alkaloids, separation of, 162
Alkanes, separation of, 159
Alkenes, separation of, 159, 162, 163
Alkyl halides, detection of, 53, 96
  separation of, 161
Amines, derivatives of, 70–74
  detection of, 96
  separation of, 162
Amino acids, derivatives of, 71–74
  separation of, 163, 164
Amino alcohols, derivatives of, 71–74
Arenes, separation of, 159
Argon detector *see* Detector, argon ionisation
Aryl halides, detection of, 53
  separation of, 161
Attenuator, 67, 68
Automatic data processing, 85, 86, 113

B

Backflush technique, 124–127
Ball and disc integrator *see* Integrator, ball and disc
Band, chromatographic *see* Peak
Baseline, drift of, 86, 119

C

Carbohydrates, derivatives of, 70–74
Carbonyl compounds, detection of, 96
  separation of, 159, 160
Carboxylic acids, derivatives of, 70–74
Carrier gas, 2, 54, 55, 65
  effect on column performance, 8, 37, 65

flow rate, measurement of, 53, 54
  leaks, 65
  pressure and flowrate, 15, 16, 65, 66
  purification of, 55
  velocity of, 8
Capillary column, 2, 35–37
  diameter of, 35, 37
  film thickness of liquid phase, 37
  performance of, 36
  sample injection system for, 23
Charles' Law, 16
Chromatographic separation, efficiency of, 7–13
  mechanism of, 4
Chromatography, history of, 1, 2
  definition of, 1
Collection procedures *see* Sample, collection of
Column, 1, 31–37
  bleeding, 44, 64, 66, 109, 119
  catalysis on, 66, 68
  conditioning, 35
  dual, 44
  effect of diameter on performance of, 10
  effect of sample overloading, 12, 35
  ideal, 6
  length, effect of, 7–9, 13, 64
  liquid phases for *see* Liquid phase
  materials for, 32
  non-ideal, 6
  open tubular *see* Capillary column
  packed, 1, 31–35
  packed capillary, 37
  packing, particle size of, 9, 33
  performance of, 64, 65
  preparation of, 34
  procedures for packing, 34
  switching procedures, 124–127
Column Efficiency, 7–13
  effect of flow programming on, 124
  effect of temperature programming on, 119

Column Temperature,
  effect on retention, 16, 66
  optimum operational, 30, 66
Company directory, 188–191
Corrected retention volume *see* Retention volume corrected
Cryogenic gas-liquid chromatography *see* Temperature programming, subambient

D

Dead volume, measurement of, 14
Decomposition of sample, detection of, 99, 112, 113
Definitions of GLC terms, 137–141
Derivatives, preparation of, 68–74
Detector, 37–53
  applications of, 52, 53
  argon ionisation, 51
  choice of carrier gas, effect of, 53
  concentration, 38, 53, 77, 78
  electron capture, 48–51
  flame ionisation *see* Flame ionisation detector
  flame photometric, 52
  gas density balance, 2, 39
  hydrogen flame temperature, 40
  ionisation, 45–52
  ionisation gauge, 41
  mass, 38, 53, 77, 78
  nitrometer, 39
  radioactive source ionisation, 48–51
  rapid scan spectrometers, 41, 99–114
  response factor, 40, 78–80
  relative response factor, 79
  selection of temperature, 42, 67
  sensitivity of, 53
  specific for organo-halogen compounds, 46, 49
  specific for organo-phosphorus compounds, 46, 52
  specific for organo-sulphur compounds, 50, 52
  thermal conductivity *see* Thermal conductivity detector
  thermionic, 46–48
  thermistor, *see* Thermal conductivity detector
  titration cell, 2, 39

Diastereoisomers, separation of, 163, 164
Digital integrator *see* Integrator, digital
Distribution coefficient *see* Partition coefficient
Distribution isotherm *see* Isotherm
Drift, correction of, 119, 148, 149
Dual column systems, 44, 119, 120

E

Eddy diffusion, 8
Electron capture detector *see* Detector, electron capture
Electron impact detector *see* Detector, thermionic
Elemental analysis, 28
Enantiomers, separation of, 163, 164
Esters, preparation of, 70, 71, 73, 74
  separation of, 160
Essential oils, separation of, 162

F

Flame ionisation detector, 45, 46
  design of, 45
  factors affecting performance of, 45, 46
  sensitivity of, 45, 53
Flowmeter, 53, 54
  calibration of, 16
Flow programming, 124
  effect upon column efficiency, 124
Flow rate of carrier gas,
  effect upon detector response, 77, 78
  measurement of, 53, 54
  optimum, 65, 66
  selection of and regulation of, 65–67
Fraction cutting, 126, 127
Functional group analysis, 95, 96

G

Gas, removal of contaminants from *see* Carrier gas
Gas density balance *see* Detector, gas density balance
Gas flow rate *see* Flow rate of carrier gas
Gas hold-up time *see* Dead volume

SUBJECT INDEX

Gas-liquid chromatogram,
  ideal distribution of peaks on, 58
  integration of, 84–88
  interpretation of, 77–116
Gas-liquid chromatograph,
  operation of, 57–74, 143–152
  typical, 21
Gas-liquid chromatograph—mass spectrometer combination, 110–113
Gas-liquid chromatography,
  basic theory, 4–18
  history of, 1, 2
  instrumentation for, 21–55
  terms used in connection with, 137–141
Gas sampling valves, 28, 29
Gases, separation of, 162
Gas-solid chromatography, 2
Gaussian peak, 8
  area of, 81–88
  relationship between standard deviation and base width of, 8
  relationship between standard deviation and height of, 82
Geometric isomers, separation of, 162, 163
Glossary of GLC symbols and abbreviations, 141, 142
Golay column *see* Capillary column

H

Halogeno-compounds,
  detection of, 53, 96
  separation of, 161
Height equivalent to a theoretical plate (HETP) *see* Plate height
Henry's Law, 4
Hydrocarbons,
  separation of, 159
  use for calibration, 15, 93–95
Hydrogen bonding, between solute and liquid phase, 59, 68, 69

I

Injection system, 22–29
  automatic, 128, 129, 131
  catalysis in, 24
  design of, 22–24
  direct on-column, 24
  encapsulated sample, 26, 27, 131
  for gases, 28, 29
  for liquids, 24–26
  for solids, 26–28
  stream splitting, 23, 36
  temperature of, 23
Integrator,
  ball and disc, 84, 85
  digital, 85–87
Ion emission detector *see* Detector, thermionic
Ionisation detector *see* Detector, ionisation
Isolation techniques *see* Sample, collection of
Isomerisation on column, 66, 68
Isotherm,
  linear, 6
  non-linear, effect upon peak shape, 6
Inert gases,
  insensitivity of the flame ionisation detector to, 46
  retention of, 14
Infrared spectroscopy, 100–105
  cavity cell for, 101
  collection of samples for, 100–105
  gas cell for, 103–105
  KBr disc technique, 102
  light pipes for, 104
  relationship between scanning speed and resolution, 104–105
  sample requirement for, 100
  use of a beam condenser, 105
Instrument directory,
  analytical instruments, 166–181
  preparative instruments, 182–187
Internal normalization technique, 79

K

Katharometer *see* Thermal conductivity detector
Ketones, separation of, 159, 160
Kovats retention index,
  for isothermal analysis, 93–95
  for programmed temperature analysis, 123

L

Leaks, detection of, 65

Liquid crystals, use as stationary phase, 163
Liquid phase, 58–65, 153–158
"bleeding" see Column bleeding
coating techniques, 33–35
effect of loading upon retention, 14
effect of polarity upon retention, 58, 59
electron donor—acceptor character of, 61, 62
film thickness of, 9, 37
loading, 64
mixed, 33, 63, 64
polarity (see also polarity index), 58–65
selection of, 33, 57, 60–64
specific uses for, 159–164
table of, 63, 153–158
temperature limits of, 63, 66, 153–164
Literature and commercial organisations, 3, 4, 18, 188–191
Longitudinal diffusion, 9

M

Mass spectrometry, 108–114
combination with GLC, 110–113
isolation of a sample for, 114
sample requirement for, 109
use in the analysis of unresolved peaks, 113–114
Mass transfer resistance, contribution to band broadening, 9
Mercaptans see Sulphur compounds
Mobile phase see Carrier gas
Molecular separators,
Becker, 111, 112
Watson and Biemann, 111, 112
Silicone membrane, 112
Multiple column systems, 63, 124–127
"flip-flop" technique, 126

N

Negative peaks, 150
Net retention volume see Retention volume, net
Non-Gaussian peaks,
area measurement of, 83–86, 88–91
Non-symmetric peaks (see also Non-Gaussian peaks), 6, 10, 12

Nuclear magnetic resonance spectrometry, 105–108
computer average transients, 107–108
isolation of samples for, 101, 105
microcells for, 106, 107
sample requirement for, 105–106

O

Open tubular column see Capillary column
Operational faults, 145–152
Operational procedure, 143, 144
Optimum carrier gas flow rate see Flow rate
Oven, 30, 31
Overlapping peaks,
automatic integration of, 87
effect on peak shapes, 12
manual integration of, 89–91

P

Packed columns, 31–35
eddy diffusion in, 8
longitudinal diffusion in, 9
performance of, 7–13, 64, 65
plate number of, 8
preparation of, 34
Particle size of solid support, 9, 33
Partition coefficients, 4, 5
Partition ratio see Partition coefficients
Peak,
area of, 77–87
broadening of, 7–13, 66
position of see Retention
relative area, 78–81
shape of, 6, 10, 12
standard deviation of, 7, 81, 83
tailing, 6, 10, 150
Peak height, 87, 88
Peak width, factors controlling, 7–13, 66
Phase ratio, 64
Phenols,
preparation of derivatives, 70–74
separation of, 160–161
Planimetry, 83, 84
Plate height, 7
equation for, 8

SUBJECT INDEX

Plate number, 8
Polarity index, 60
Polarity of liquid phases *see* Polarity index and Liquid phases, polarity
Process GLC, 130–132
Preparative GLC, 127–130
  columns for, 127, 128
  operational cycle, 129, 130
  sample introduction, 128, 129
  traps for, 130
Programmed GLC,
  temperature programmed, 117–123
  flow programmed, 124
Pyrolysis GLC, 27, 28

Q

Qualitative analysis, 91–114
  functional group analysis, 95, 96
  use of different liquid phases, 92, 93
  use of IR spectroscopy, 100–105
  use of Kovats retention index, 93, 94
  use of mass spectrometry, 108–114
  use of NMR spectrometry, 105–108
  use of retention index data, 91–95
  use of thin layer chromatography, 96–99
  use of UV spectrometry, 99
Quantitative analysis, 77–91
  area measurement, 81–87
  calibration curves, 80
  internal normalisation, 79
  internal standard techniques, 79
  measurement of peak height, 87, 88
  use of reference standards, 79, 80

R

Radioactivity—GLC system, 176
Recorder,
  peak attenuation, 67, 68
Relative retention, 17
Retention of chromatographic peaks, 13–18
  effect of column temperature on, 66
  effect of temperature programming on, 120–123
  of homologous series, 14, 58, 59
  relationship with boiling point of sample, 58, 59

Retention index *see* Kovats retention index
Retention measurements, reference standards for, 17
Retention temperature, 120–123
  effect of heating rate on, 121–123
  effect of initial column temperature on, 120–122
Retention time, 13
  effect of column temperature on, 66
  effect of flow rate on, 13, 66
Retention volume, 13
  adjusted, 14
  corrected, 15
  effect of column length on, 13
  effect of column temperature on, 66
  factors controlling, 13, 14, 66
  measurement of, 14, 15
  net, 15
  obtained from columns of differing polarity, 60–64
  of inert gases, 14
  relationship with carbon number, 14, 94
  specific, 16
Resolution of chromatographic peaks, 11–13, 35
Response factor *see* Detector response factor

S

Sample,
  calculation of percentage composition, 78
  collection of, 101, 102, 130
  encapsulated, 26, 27, 131
  enthalpy of vaporisation of, 16
  input distribution of, 10, 23
  polarity of, 61
  pretreatment of, 68–74, 95, 96
Sample introduction *see* Injection system
Sample size, 10, 23, 24
  effect upon resolution, 12, 67
  effect upon retention, 10, 11
Sample valves, constant volume, 28, 29
Sensitivity of detector, 53
Skewed peaks, area measurement of, 88, 89

Solid support, 31, 32
  catalysis on, 32
  coating of, 33–35, 63
  ideal features of, 31
  in situ coating of, 33, 63
  particle size of, 9, 33
  selection of, 31, 32
  silylation of, 32
Solids, injector for *see* Injection system
Solute (see also Sample), 140
  relationship between boiling point and retention, 60
Solvent flush technique, 25
Specific retention volume *see* Retention volume, specific
Spectroscopic analysis of GLC effluent, 100–114
Stationary phase (see also Liquid phase), 1
Steroids, separation of, 162
Stream splitting, 21, 23, 36, 100, 112
Structure determination, by the use of Kovats retention index, 94, 95
Sulphur compounds, detection of, 53, 96
  separation of, 161, 162
Survey run, 57–68
Syringe,
  errors in delivery from, 24–26
  sample introduction from, 24–26, 28, 79, 80

T

Tailing *see* Peak, tailing,
Temperature,
  control of, 30, 31
  effect upon column performance, 9, 10, 12, 66
  limits for liquid phase, 63, 66, 153–158
  of column, 9, 10, 66, 67
  of detector, 42, 67
  of injection system, 9, 10, 22
Temperature programmed GLC, 117–123
  calculation of Kovats retention index, 123
  effect upon column performance, 119

  for samples of widely differing polarity, 117
  for samples of widely differing volatility, 117
  linear, 118
  non-linear, 118
  retention temperature, 120–123
  subambient, 123
  techniques for, 118, 119
  use of dual columns, 119, 120
Terpenes,
  carbon skeletal analysis of, 95
  separation of, 162
Theoretical plate, 7
Thermal conductivity detector, 2, 42–45
  choice of carrier gas for use with, 42, 44
  effect of temperature on, 43, 44
  factors affecting performance of, 42–45
  filament current of, 43
  flow sensitivity of, 44
  gas circuits for use with, 44
  micro, 43
  thermistors, 44
Thermal degradation *see* Pyrolysis GLC
Thermistor *see* Thermal conductivity detector
Thin layer chromatography, 96–98
  direct coupling with GLC, 98
Total gas hold up *see* Dead volume
Total peak area measurement, 83–87
Trapping *see* Sample, collection of
Trimethylsilylation of sample, procedures for, 71–73

U

Ultraviolet Spectroscopy, 99
Unresolved peaks, 151
  area measurement of, 89–91
  separation by mass spectrometry, 113, 114

V

van Deemeter equation, 8, 36, 65